1
The World About Us

Name _____

Explore

See what you can do with .

How are  alike or the same?

How are  different?

same	different

Name _____

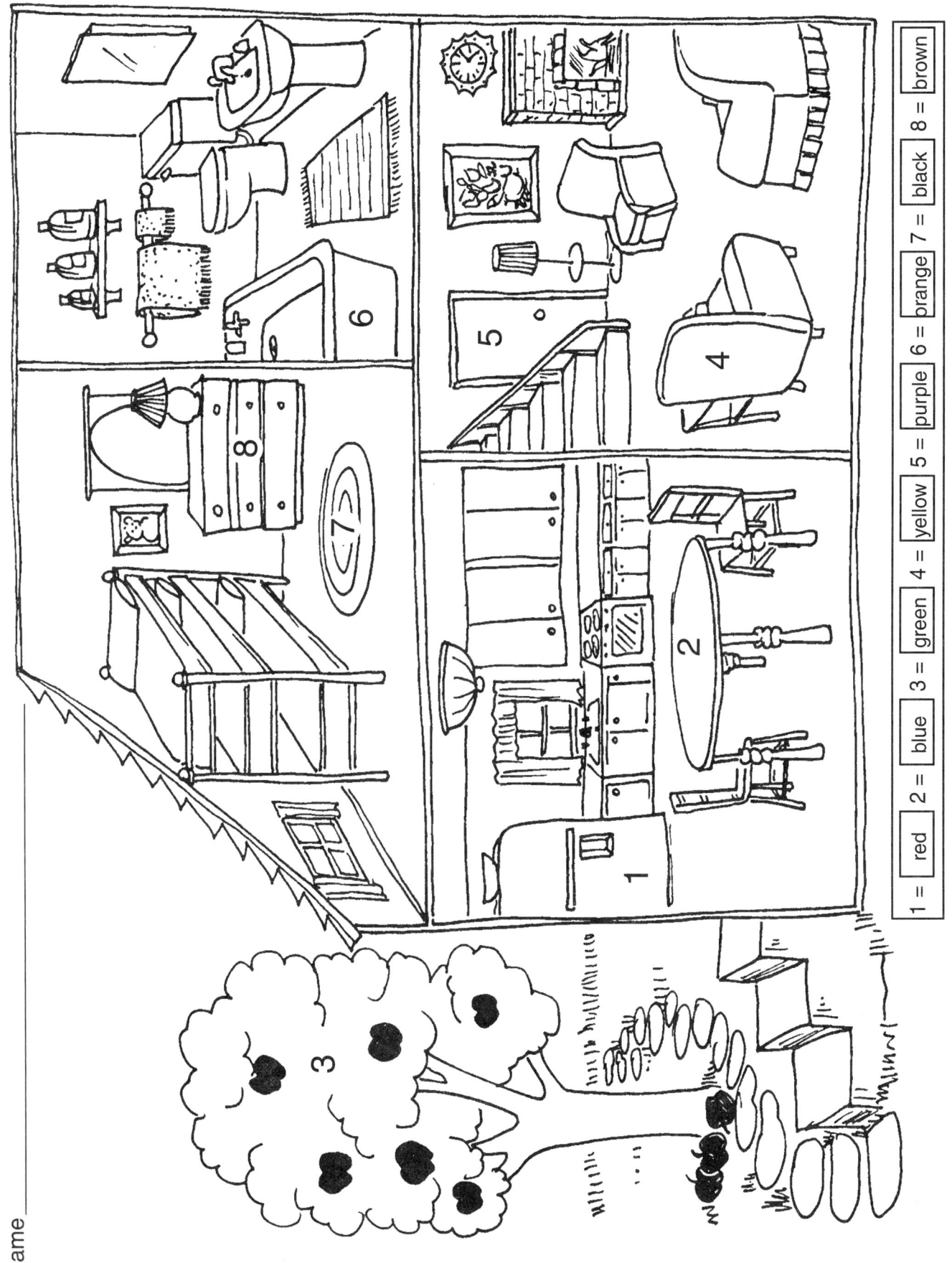

1 = red 2 = blue 3 = green 4 = yellow 5 = purple 6 = orange 7 = black 8 = brown

Name _____

Comparing Size

The elves are small or little.	The giant is big or large.

Order from smallest to largest.

Name _____

Patterns
Cover with a pattern.

Which bear comes next?

Name _____

As Many As ?
Draw lines to match.

 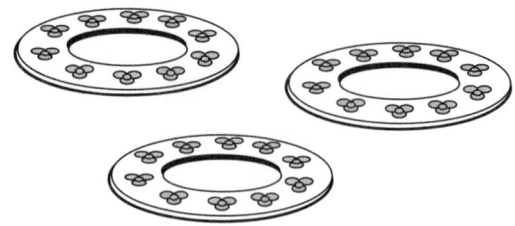 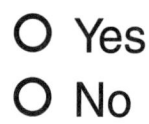

    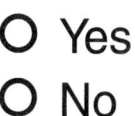

   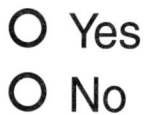

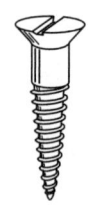 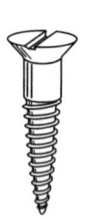 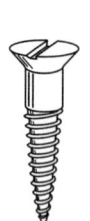 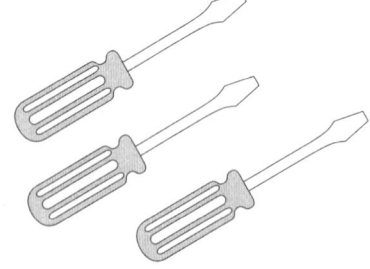

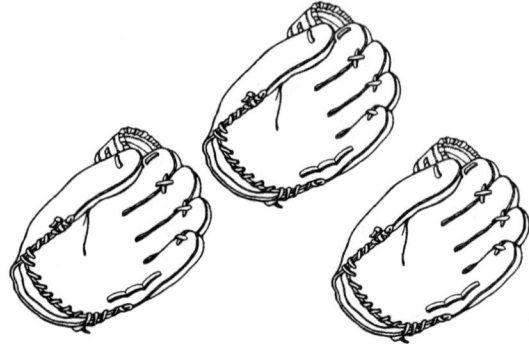   

Name _____

Bears on an Airplane

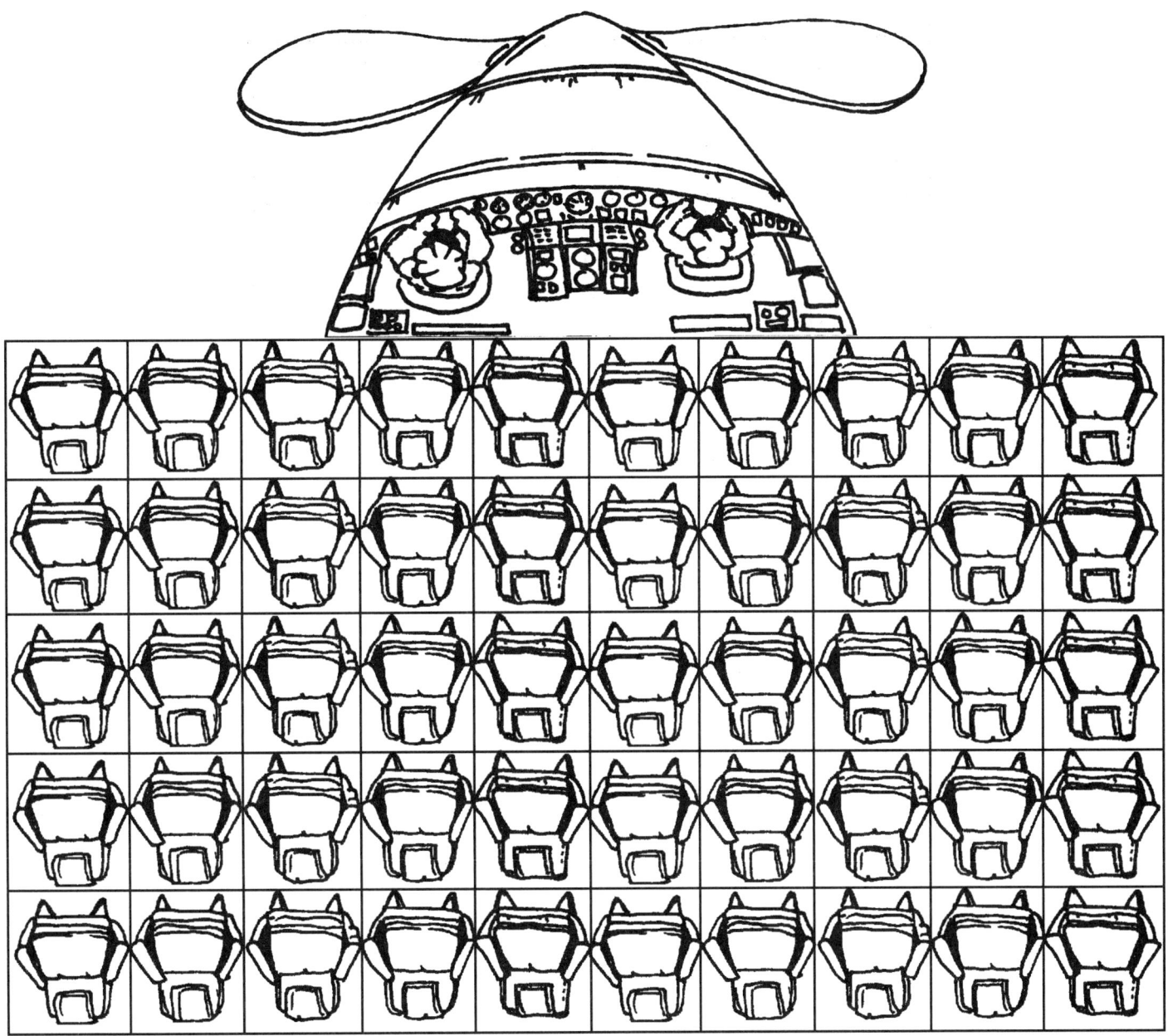

How many bears? _____ Color one seat for each bear.

How many empty seats? _____

Name _____

Our Favorite Colors

								pink
								brown
								black
								orange
								purple
								yellow
								green
								blue
								red

Name _____

Totem Poles: Top, Middle, Bottom

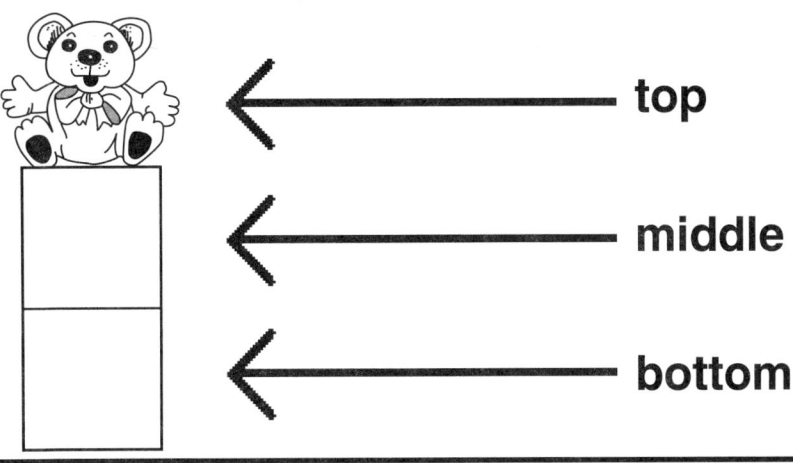

Build a pole to match. Color.

| Green |
| White |
| Black |

| Red |
| Blue |
| Yellow |

| Brown |
| Orange |
| Yellow |

top—red
middle—white
bottom—green

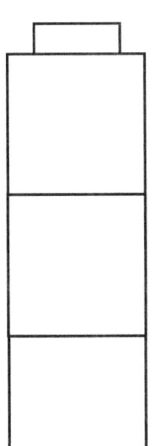

middle—green
bottom—black
top—orange

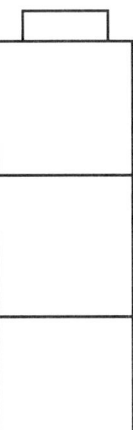

Name _____

One More

Build a set with one more.
Draw a picture.

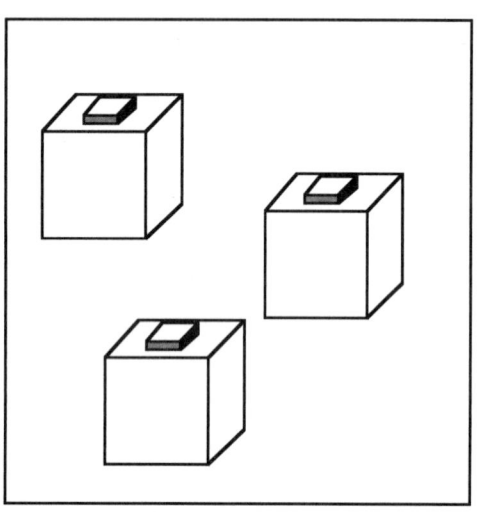

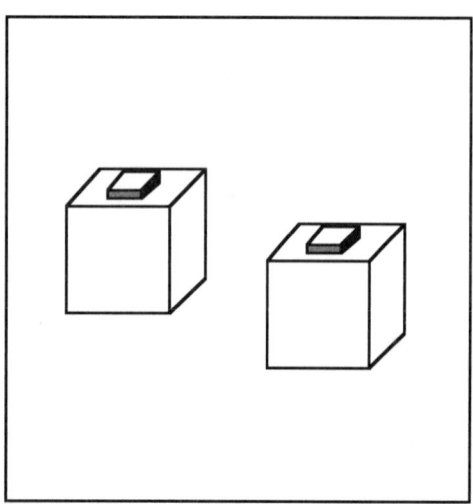

Name _____

One Less

Build a set with one less.
Draw a picture.

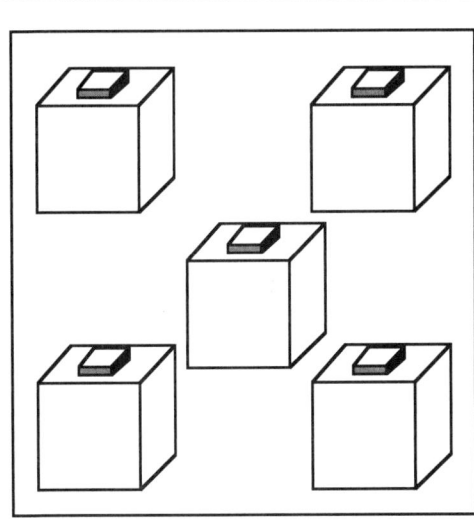

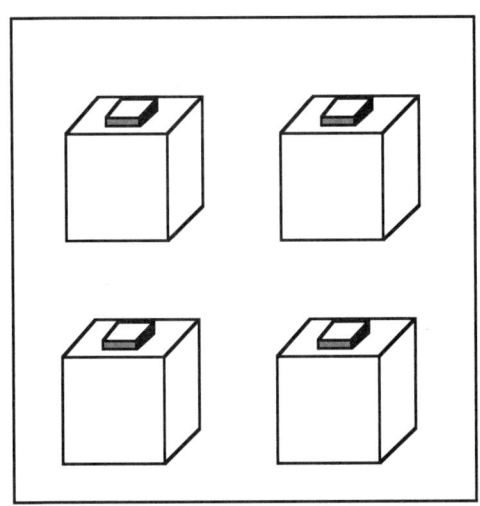

Name _____

Patterns

Use cubes of two colors.
Make this pattern. ✎ the pattern.

Make this pattern. ✎ the pattern.

Use cubes of three colors.
Make a pattern. ✎ the pattern.

Make a different pattern. ✎ the pattern

Name _____

Tallest, shortest

shortest—green

tallest—brown

shortest—yellow   next shortest—orange   tallest—brown

tallest—green   shortest—black   next shortest—brown

13

Name _____

Long, Short, Wide

The worm is short. The snake is long.

Ring longest.	Ring shortest.

Ring wider.	Ring widest.

14
© Math Teachers Press, Inc. Reproduction by any means is strictly prohibited.

Name _____

Ordering the Day

| first | next | last |

○ first   ✓ next   __ last

15
© Math Teachers Press, Inc. Reproduction by any means is strictly prohibited.

Name _____

September

Sunday	Monday	Tuesday	Wednesday	Thursday	Friday	Saturday
1	2	3	4	5	6	7
8	9	10	11	12	13	14
15	16	17	18	19	20	21
22	23	24	25	26	27	28
29	30					

How many days in all are in September?

28 ○   29 ○   30 ○   31 ○

What day of the week is September 1?

Tuesday ○   Friday ○   Monday ○   Sunday ○

How many Saturdays are in September?

3 ○   4 ○   5 ○   6 ○

What day of the week is September 30?

Sunday ○   Wednesday ○   Monday ○   Friday ○

Birthday Months

Ring the month with the most birthdays. **X** the month with the least birthdays.

	Jan.	Feb.	March	April	May	June	July	Aug.	Sept.	Oct.	Nov.	Dec.
9												
8												
7												
6												
5												
4												
3												
2												
1												
0												

Name ____________________

Curved, Straight, Open, Closed

straight lines	curved lines
open figures	closed figures

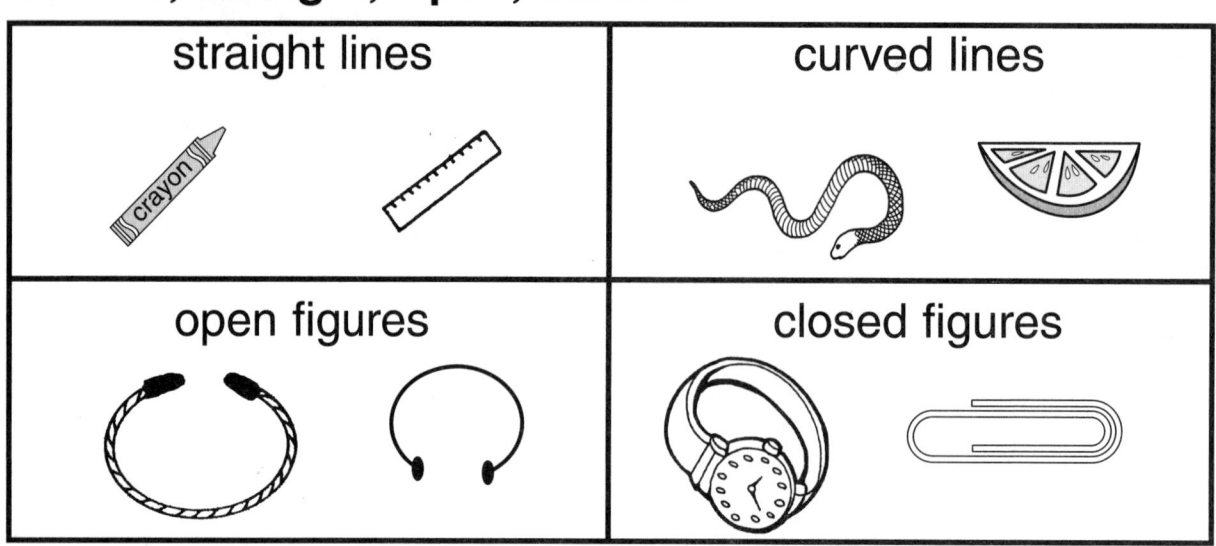

Trace straight lines green. Trace curved lines red.

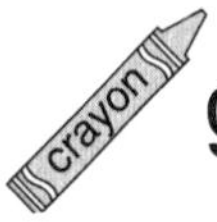

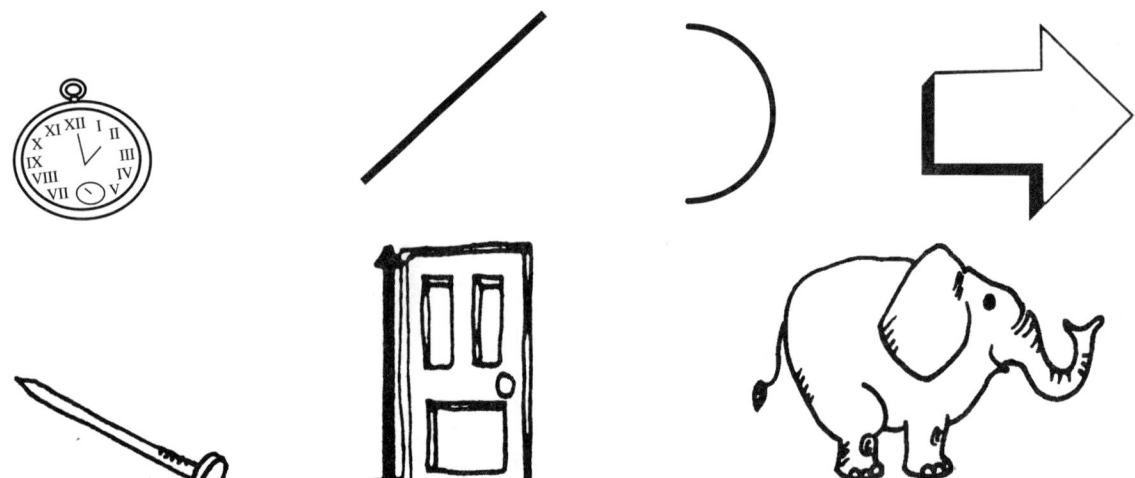

Ring each closed figure. Trace open figures blue.

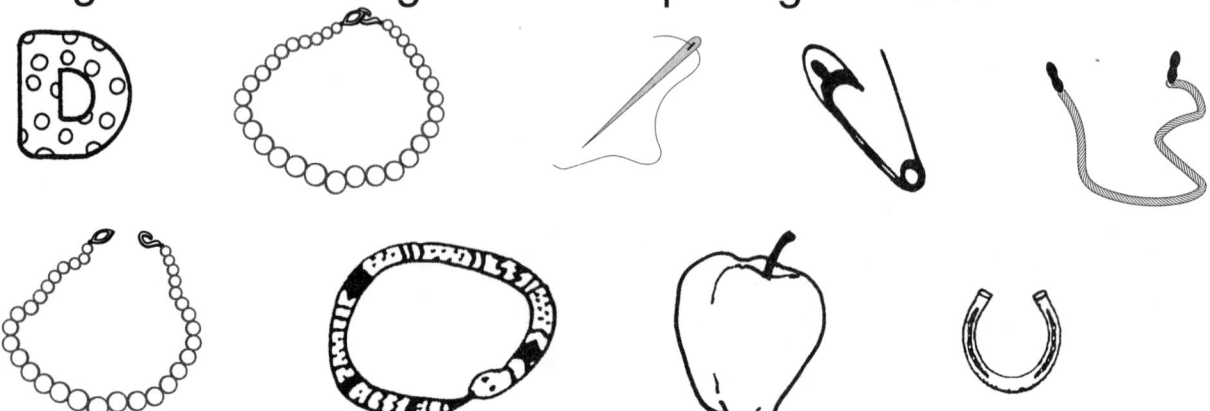

Name _____

 these shapes.

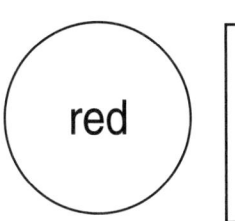 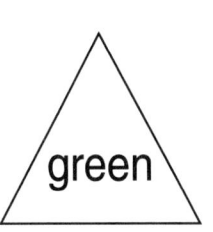 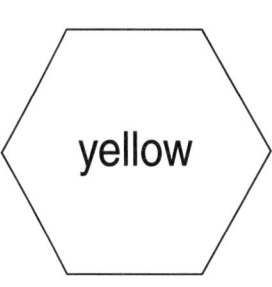

---

 the same shapes the same color.

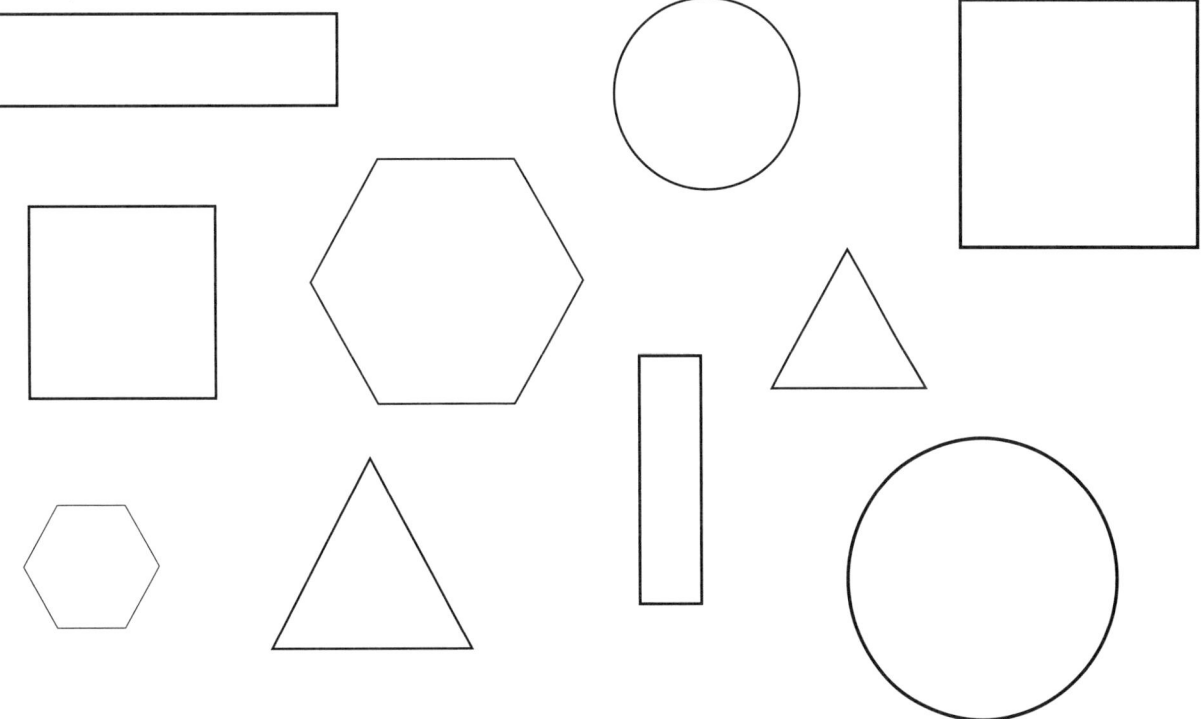

19

Name _____

Circles

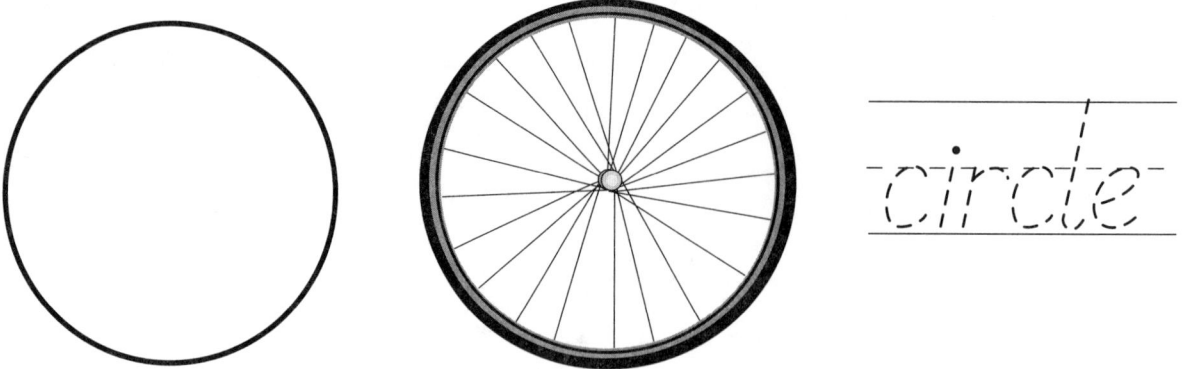

Connect the dots to make a circle.

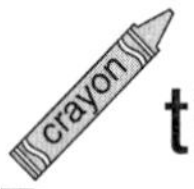

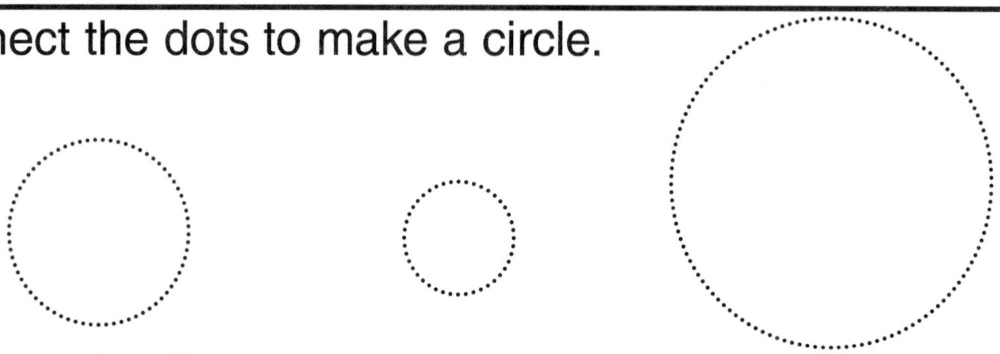

 the circles yellow.
Draw a dot inside the **small** circles.

Ring the circle.

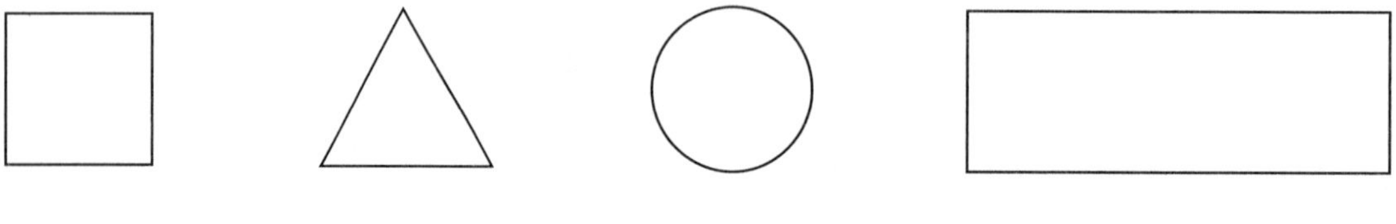

  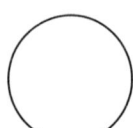 

Name _____

Squares

  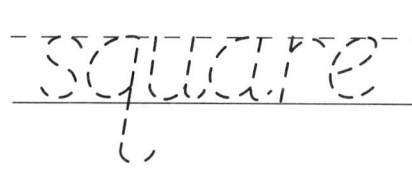

Connect the dots to draw a square.

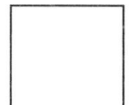 the squares purple.
Draw a dot inside each small square.

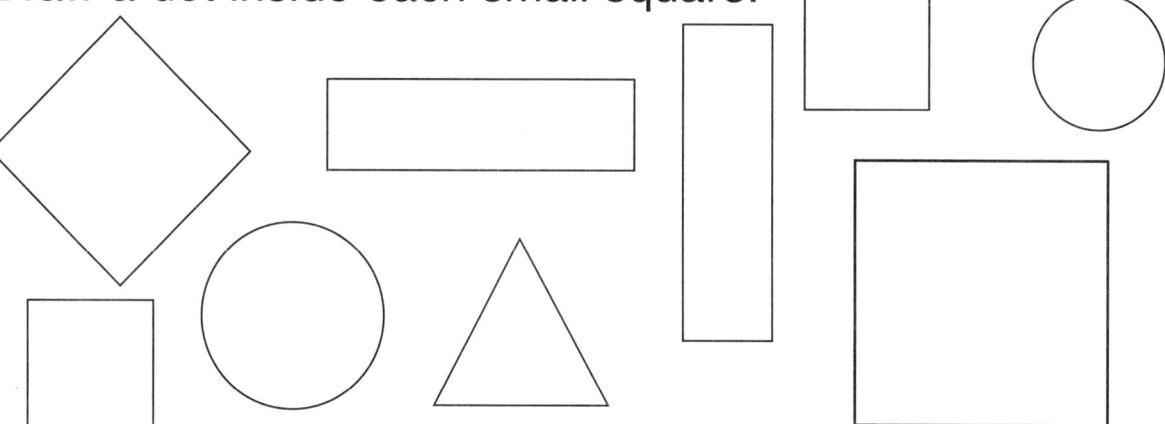

Ring the square.

  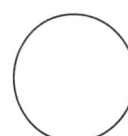

Name _____

Triangles

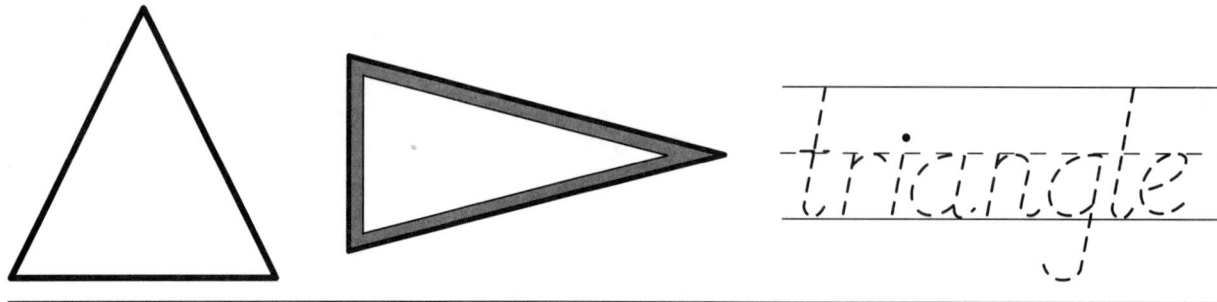

Connect the dots to make a triangle.

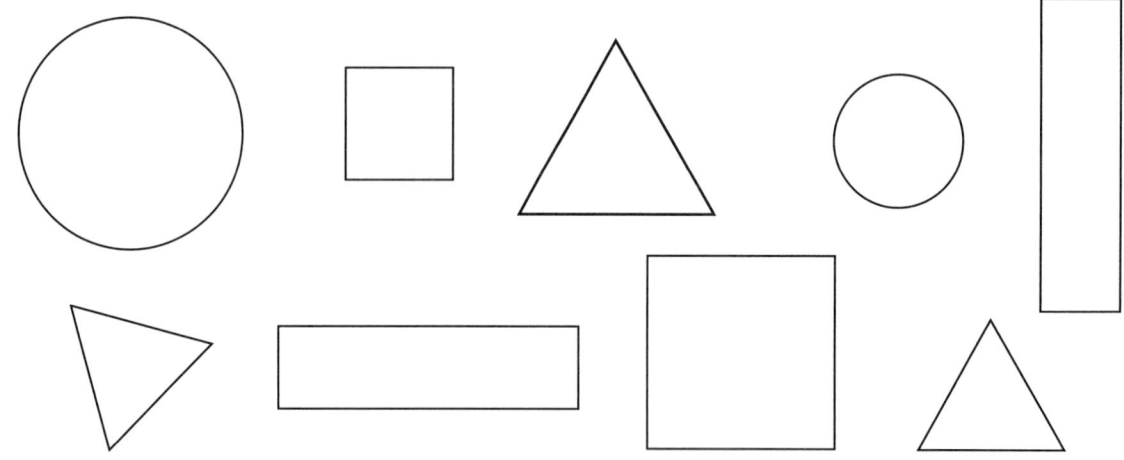

the triangles orange.
Draw a dot inside each small triangle.

Ring the triangle.

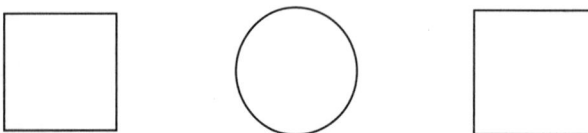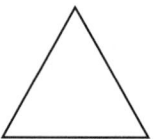

Name _____

Rectangles

  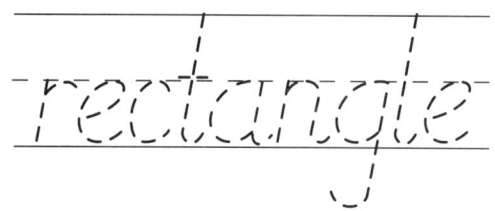

Connect the dots to make a rectangle.

🖍 the rectangles green.
Draw a dot inside each small rectangle.

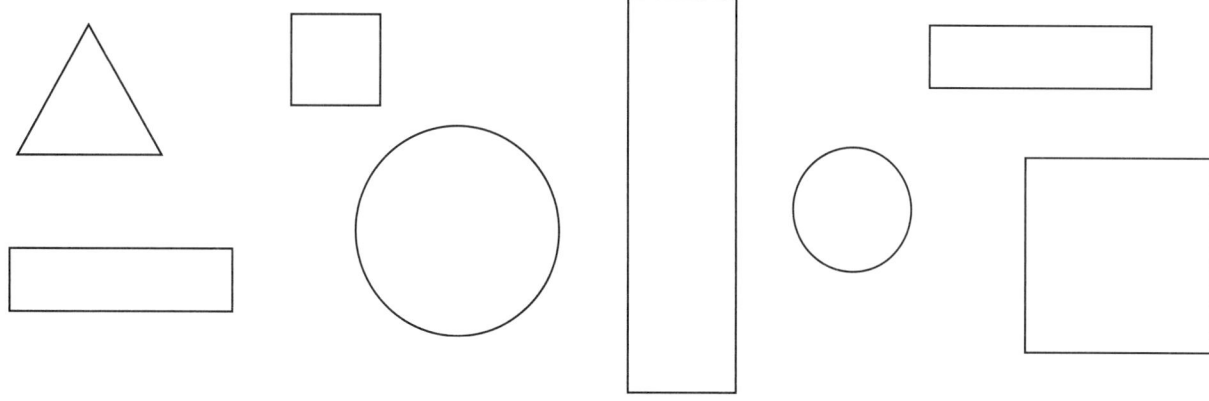

Ring the rectangle.

  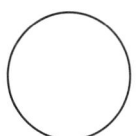 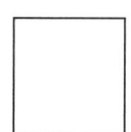

23

Name _____

| **Thick** | **Thin** |

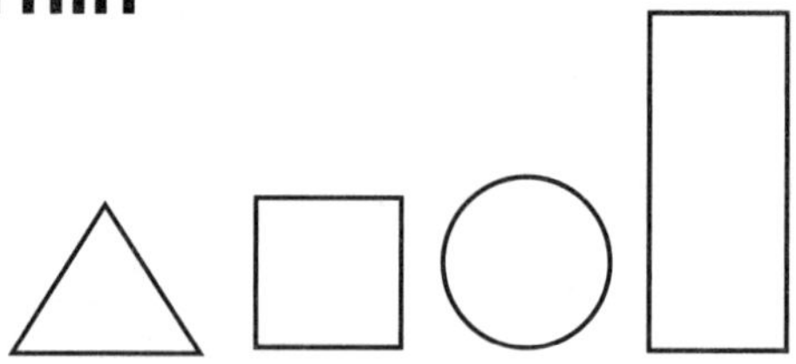

Ring the thick board. Ring the thin pizza.

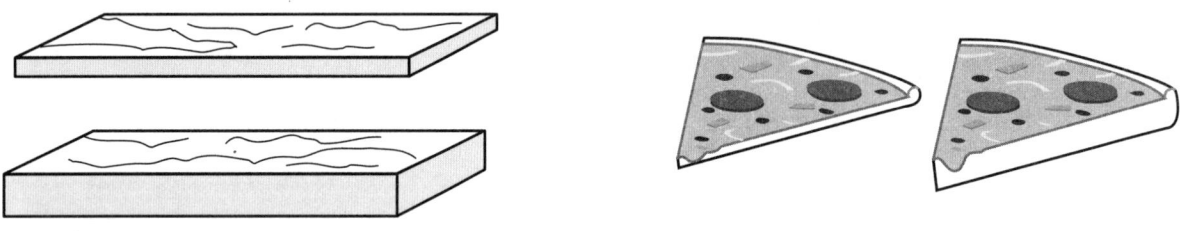

🖍 the thick shapes yellow.

🖍 the thin shapes blue.

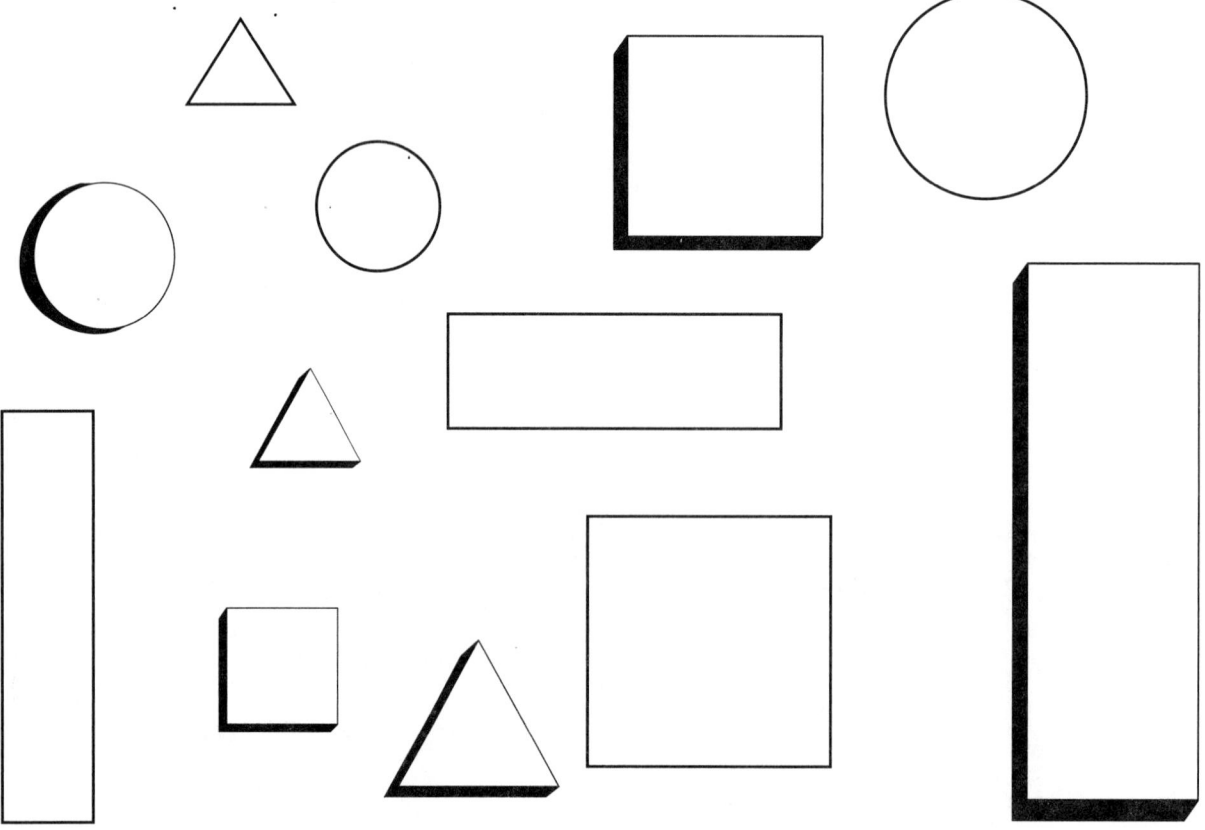

Name _____

Ovals

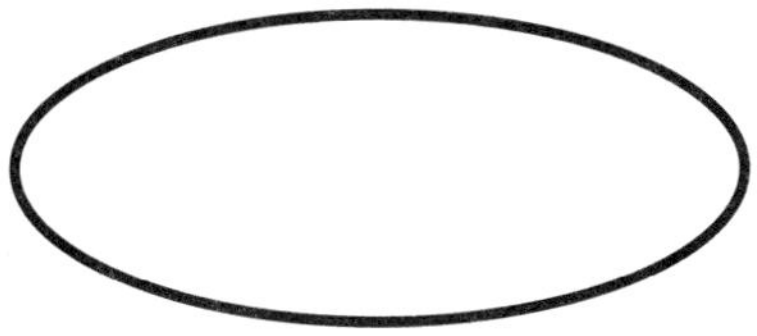

🖍 the ovals blue.
Connect the ovals.

25
© Math Teachers Press, Inc. Reproduction by any means is strictly prohibited.

Name _____

Find the pattern. Which comes next?

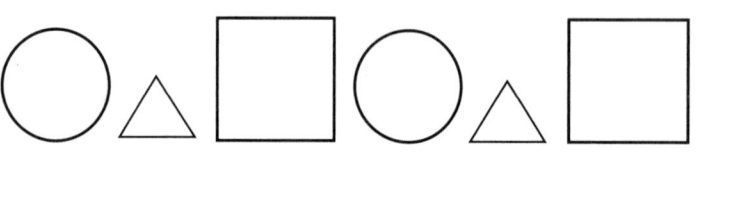

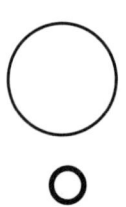

Name _____

Cubes and Spheres

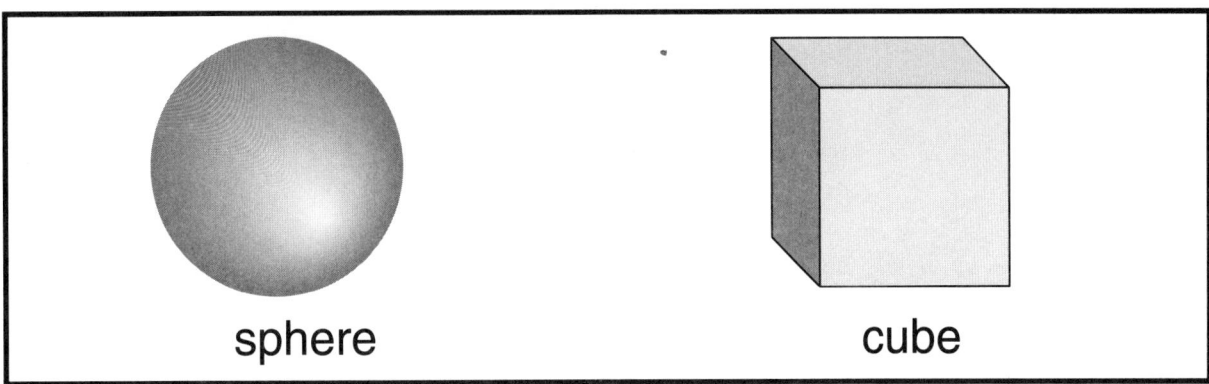

Ring.

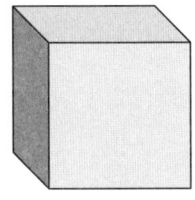

This is a   cube
            sphere

Number of sides: 4   5   6

It   can   roll.
     cannot

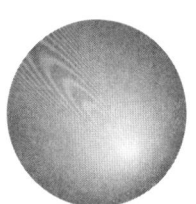

This is a   cube
            sphere

Number of sides: 0   1   2

It   can   roll.
     cannot

Ring the spheres. X the cubes.

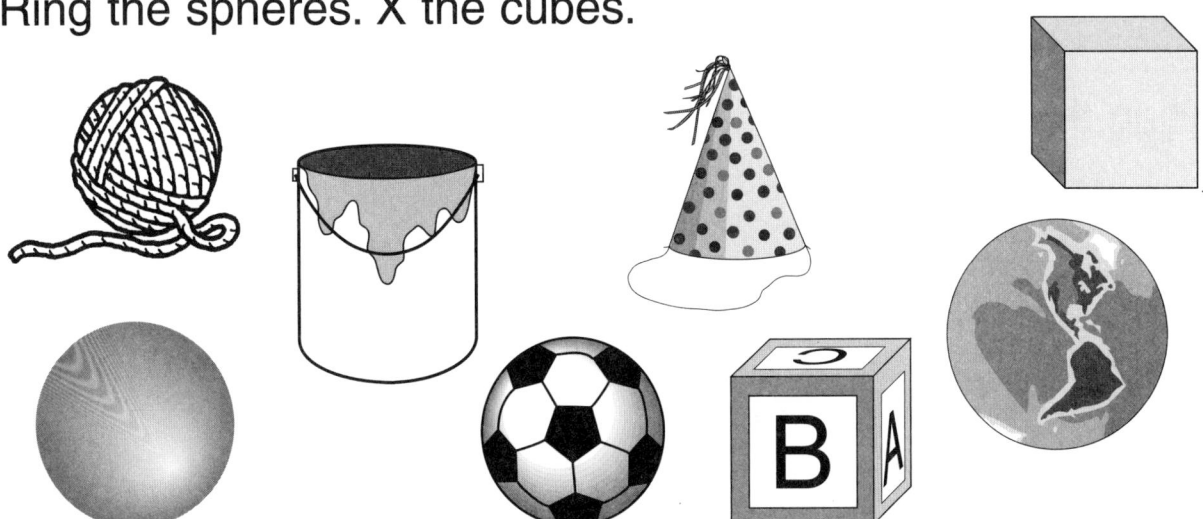

Name _____

Cylinders and Cones

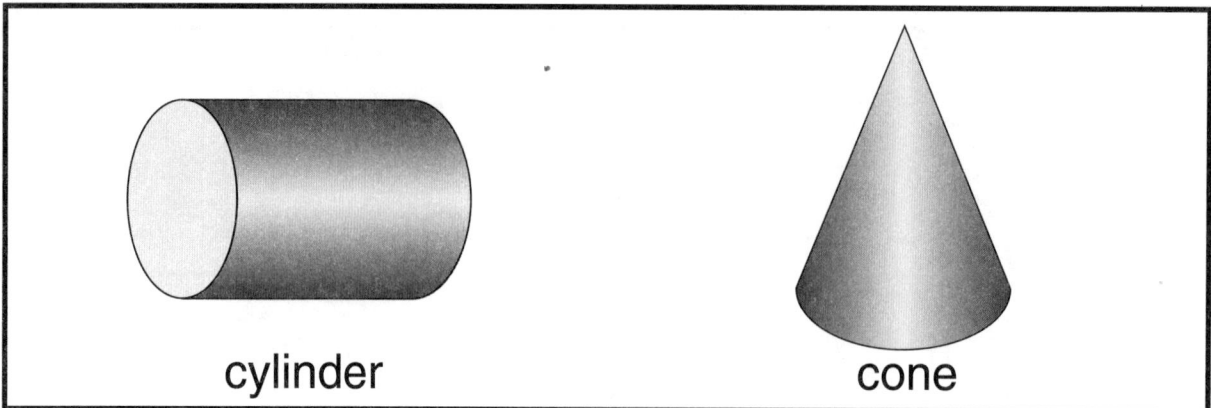

Ring.

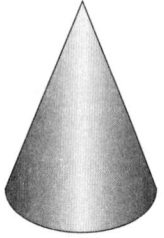

This is a   cone
            cylinder

Number of flat sides: 0 1 2

It can   roll.
    cannot

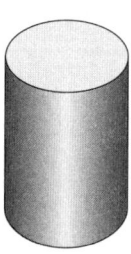

This is a   cone
            cylinder

Number of flat sides: 0 1 2

It can   roll.
    cannot

Ring the cylinders. ✓ the cones.

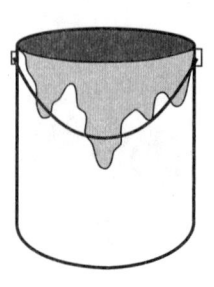   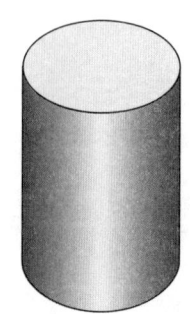

28
© Math Teachers Press, Inc. Reproduction by any means is strictly prohibited.

Name _____

What shape is on the bottom?

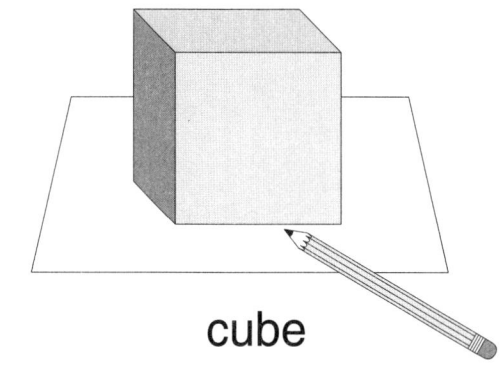
cube

square

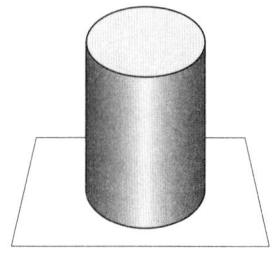
cylinder

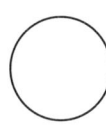   

1 O 2 O 3 O 4 O

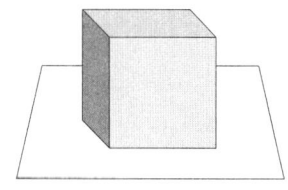
cube

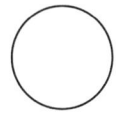  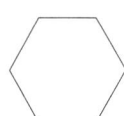 

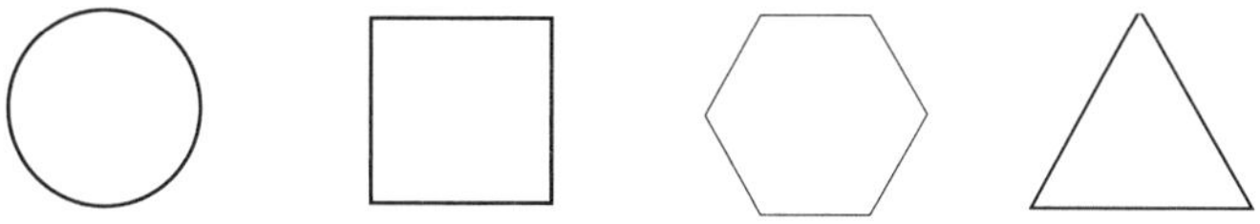

1 O 2 O 3 O 4 O

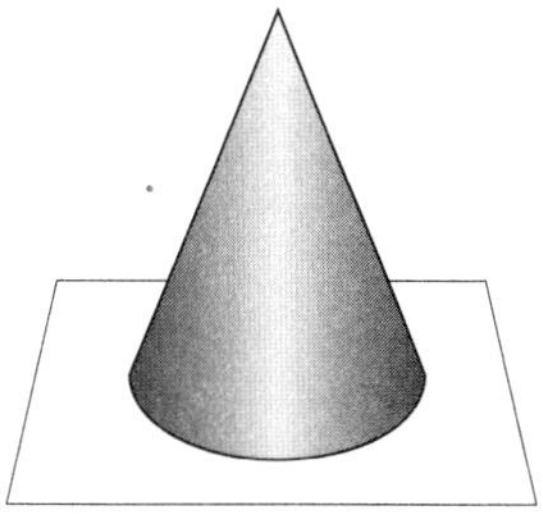

---

cone

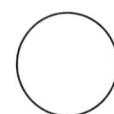 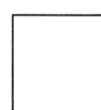

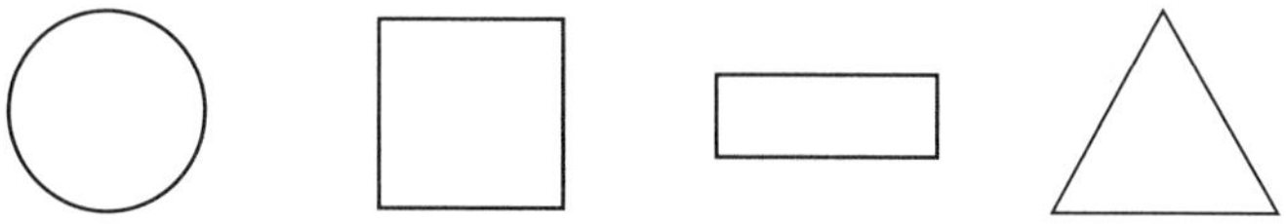

1 O 2 O 3 O 4 O

29
© Math Teachers Press, Inc. Reproduction by any means is strictly prohibited.

Name _____

Graphing Shapes

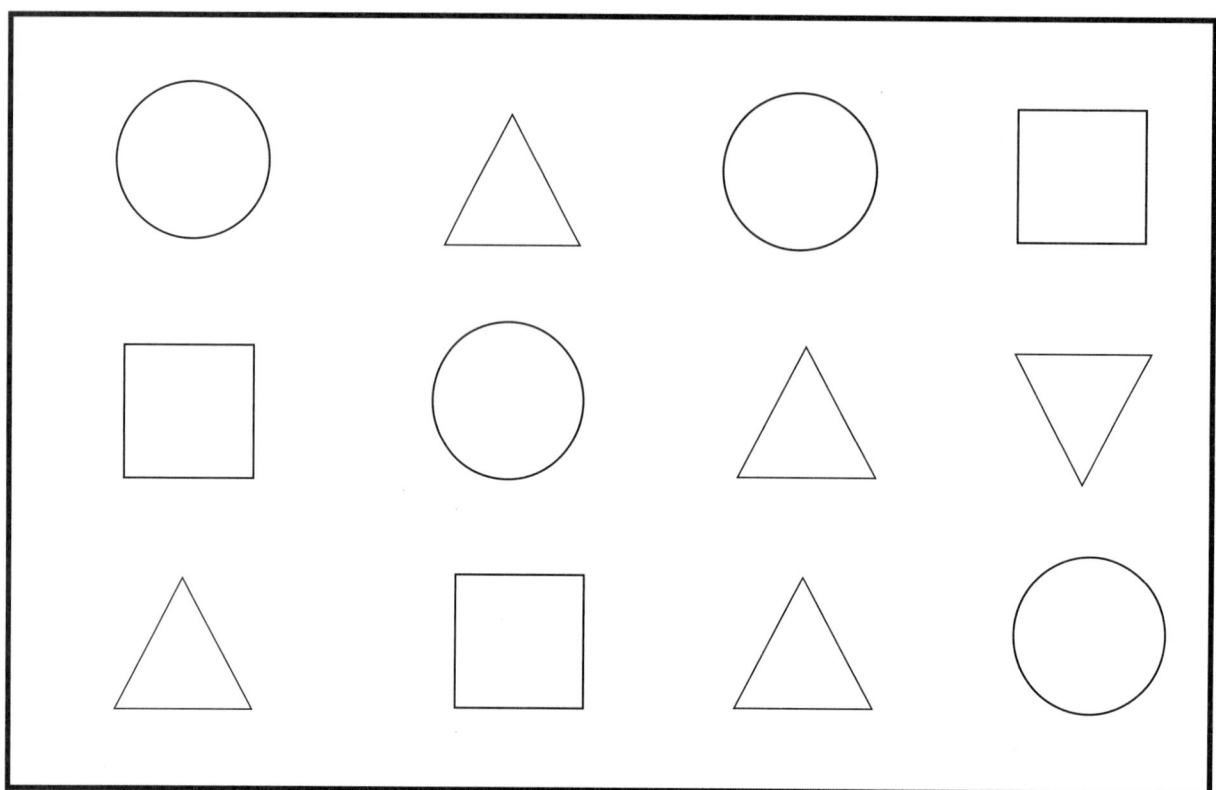

Which is a graph of these shapes?

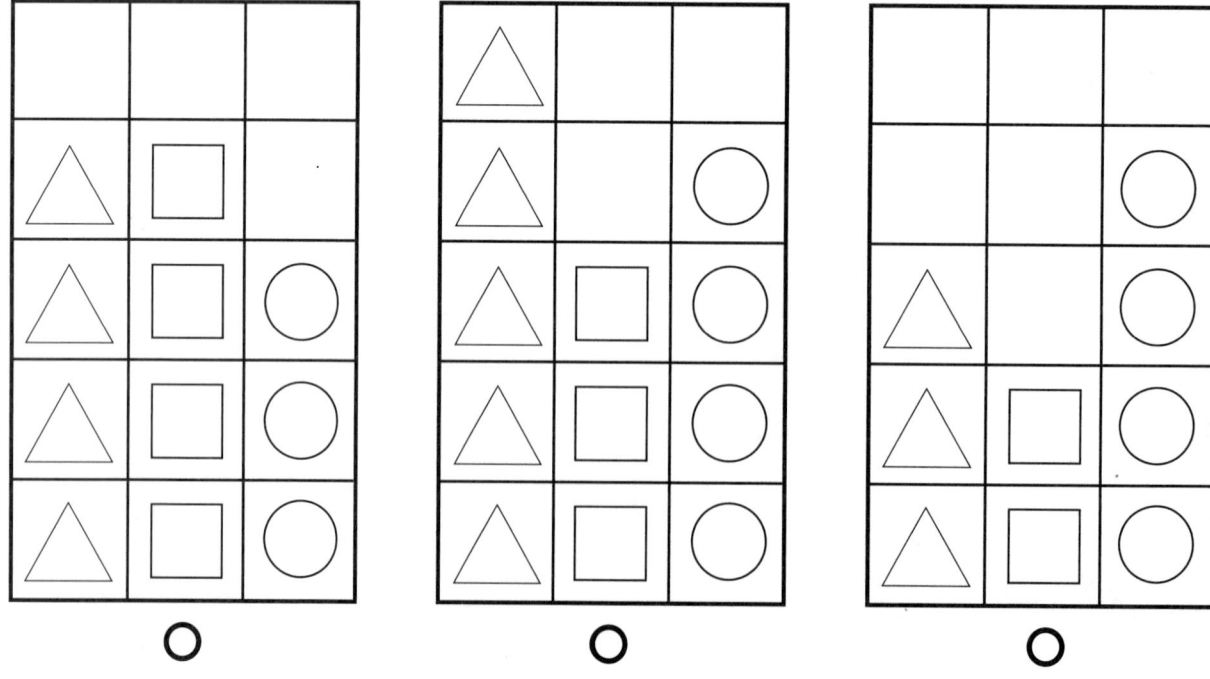

30
© Math Teachers Press, Inc. Reproduction by any means is strictly prohibited.

Name _____

Chapter 1 Review

1. Mark **X** on the one that is different.

2. Ring the shortest pencil.

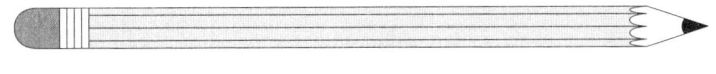
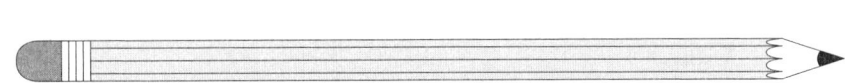

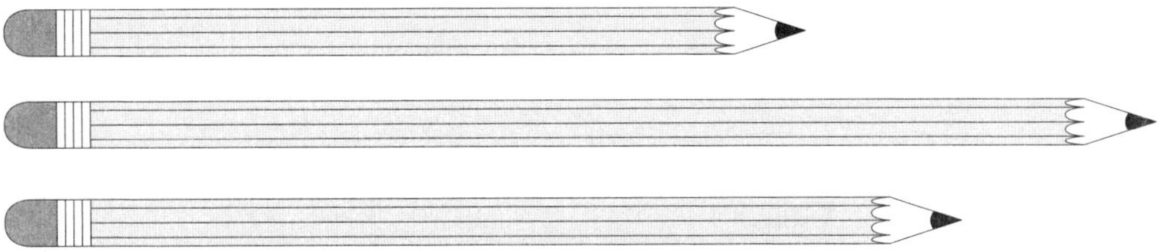

3. Find the pattern. Which comes next?

4. Ring the picture that happened first.

5. Shade the top cube.

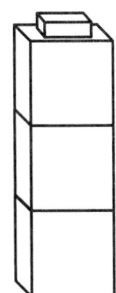

6. Ring the set that has more.

31
© Math Teachers Press, Inc. Reproduction by any means is strictly prohibited.

Name _____

Chapter 1 Review

7. Ring the square.

   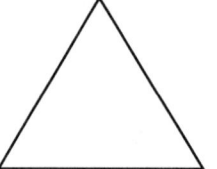

8. Ring the cube.

 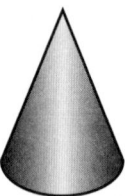 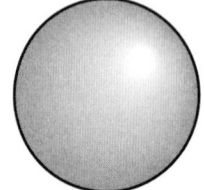 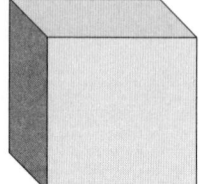 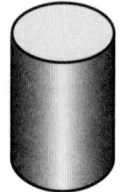

9. Which shape can be traced?

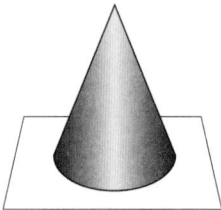  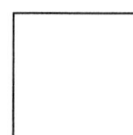 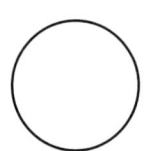

10. Which is a graph of these pictures?

32
© Math Teachers Press, Inc. Reproduction by any means is strictly prohibited.

2
Numbers to 12

Name _____

Ring the group that has the same number.

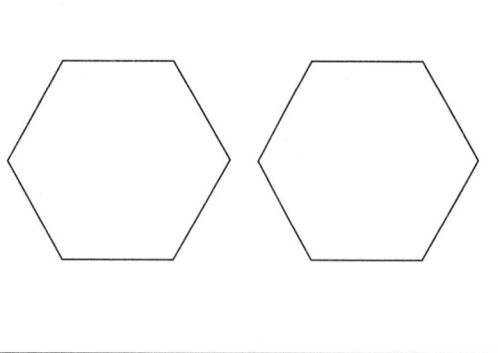 

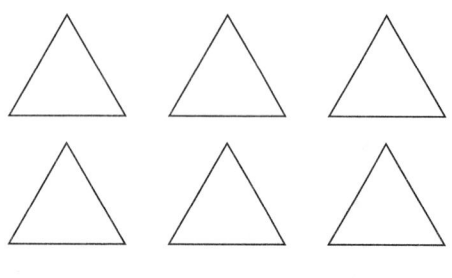 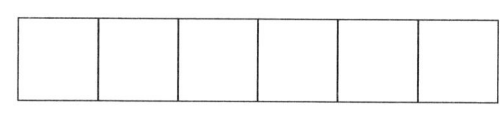

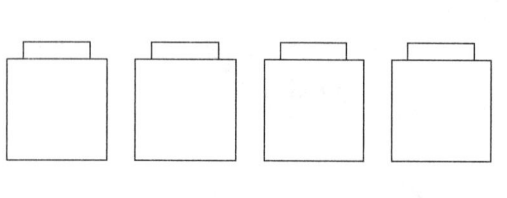 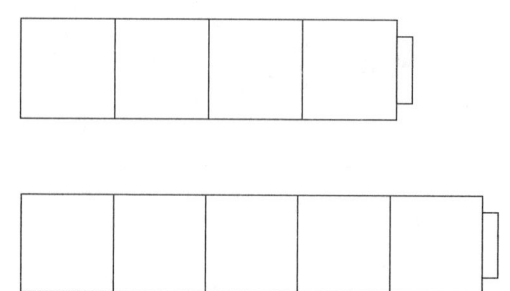

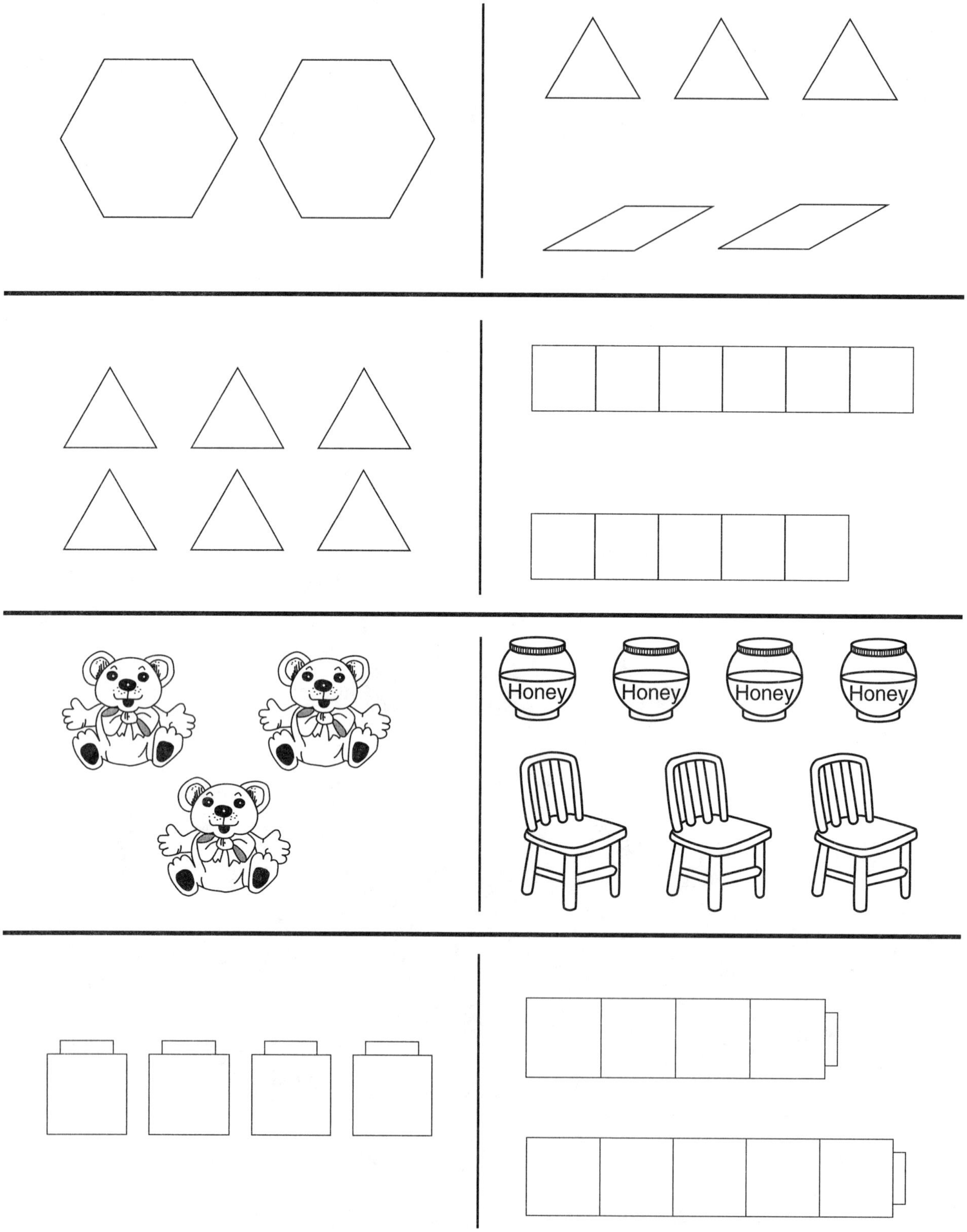

34
© Math Teachers Press, Inc., Reproduction by any means is strictly prohibited.

Name _____

One and Two

1
one

2
two

How many? Ring the number.

1   2

1   2

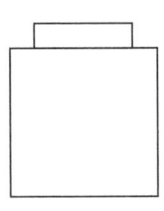 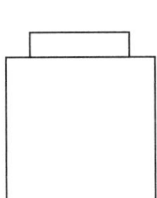

1   2

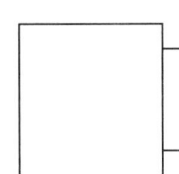

1   2

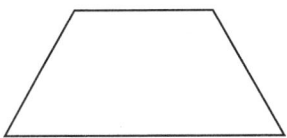

1   2

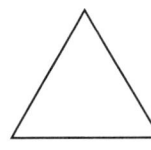 

1   2

Name _____

Write the number.

Write how many.

Name _____

Three and Four

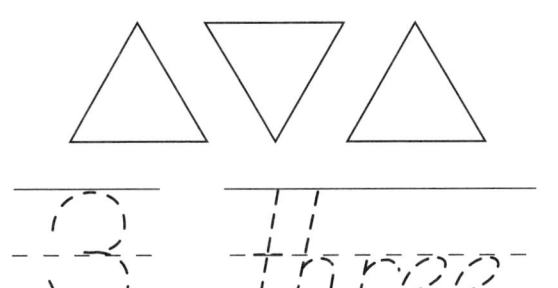

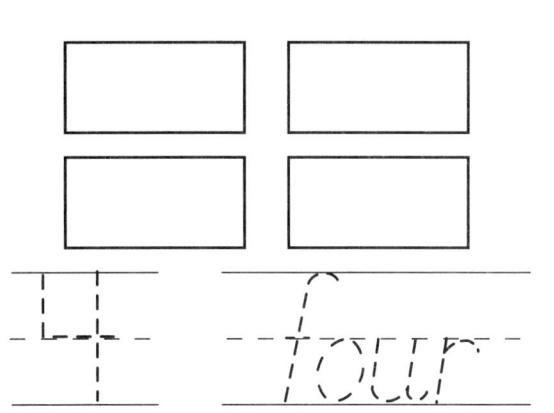

3   three   4   four

How many? Ring the number.

2 3 4	2 3 4
2 3 4	2 3 4
2 3 4	2 3 4

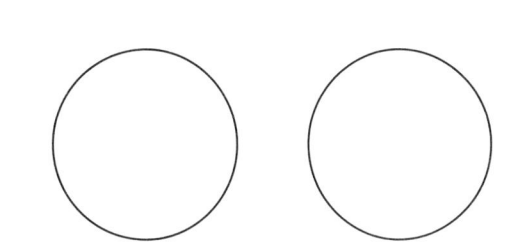

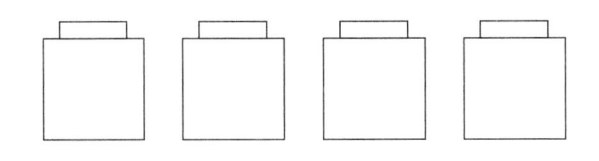

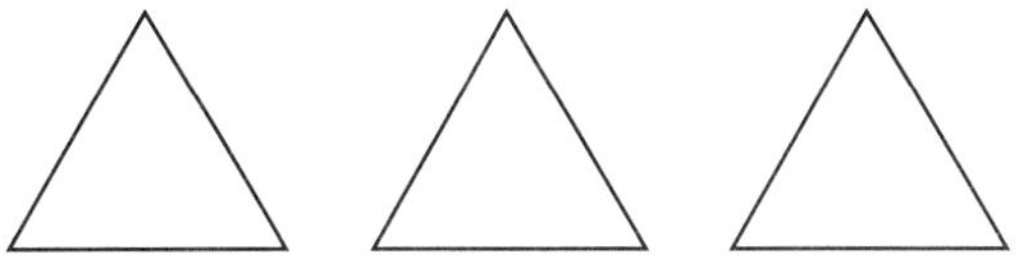

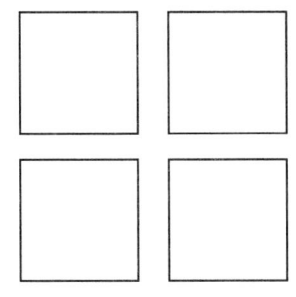

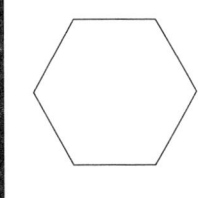

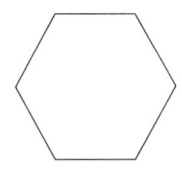

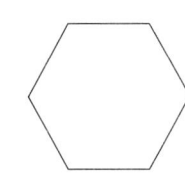

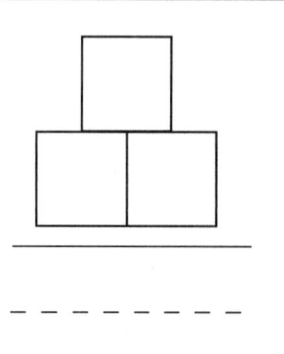

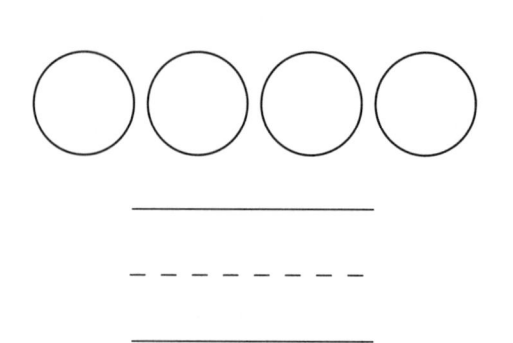

Name _____

Write the number.

Write how many.

Name _____

Zero

How many 🐭 in the lake?

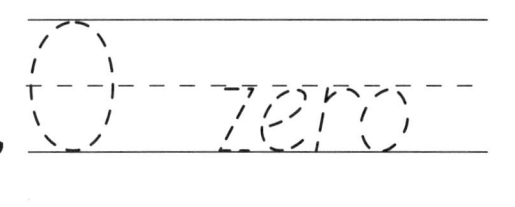

0 — zero

How many 🐭? Ring the number.

0   1   2   3 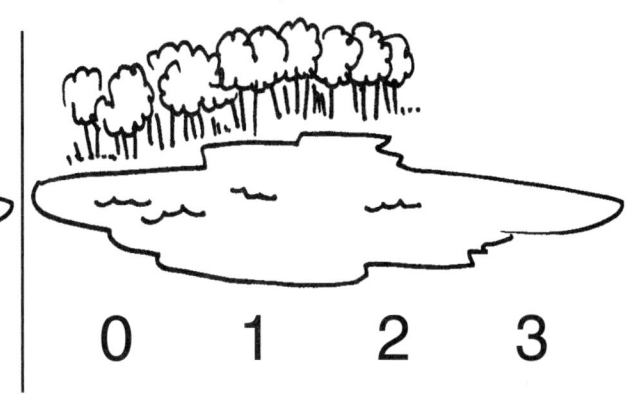   0   1   2   3

How many 🦴?

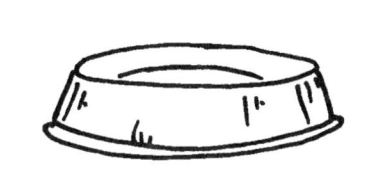

0   1   2   3    0   1   2   3

How many 🌼?

0   1   2   3    0   1   2   3

Name _____

How many 🟦 on the 🪑 ?

How many ✈ ? Write the number.

How many 🟦 ?

How many 🍌 ?

40
© Math Teachers Press, Inc., Reproduction by any means is strictly prohibited.

Name _____

Five and Six

How many ☐s?

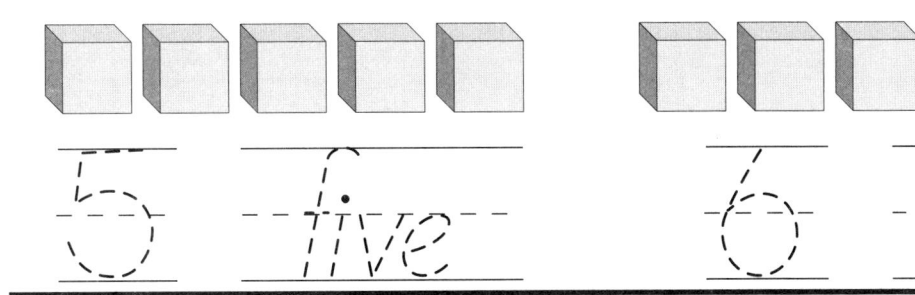

5   five   6   six

How many? Ring the number.

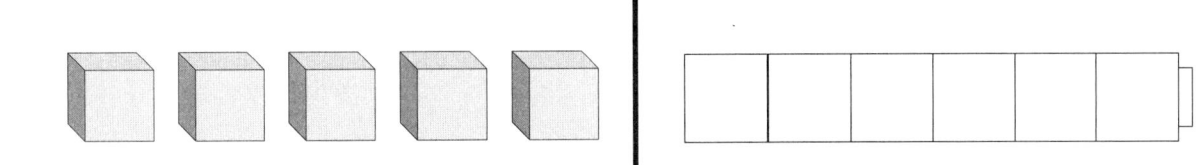

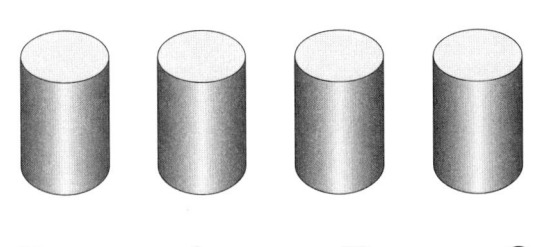

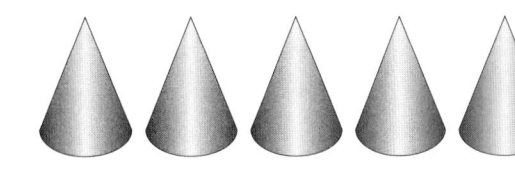

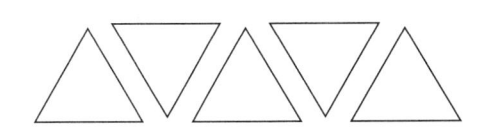

3   4   5   6 | 3   4   5   6

3   4   5   6 | 3   4   5   6

3   4   5   6 | 3   4   5   6

Name _____

Write the number.

How many? Write the number.

How many? Use 2 🖍 to color the ▢s.

42
© Math Teachers Press, Inc., Reproduction by any means is strictly prohibited.

Name _____

Seven and Eight

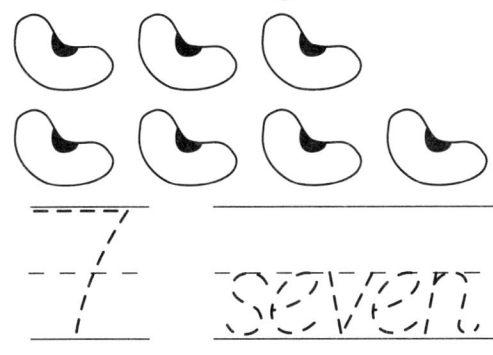

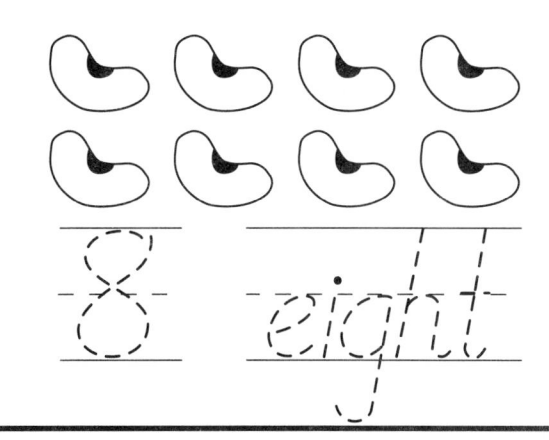

7 seven   8 eight

How many? Ring the number.

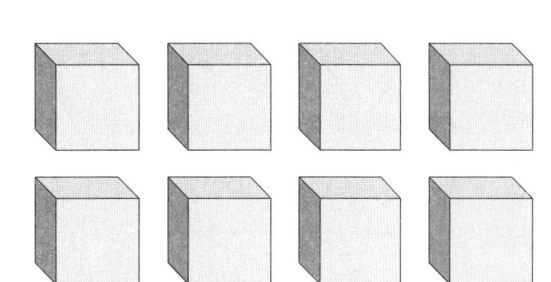

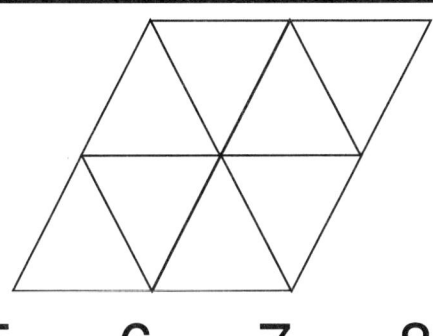

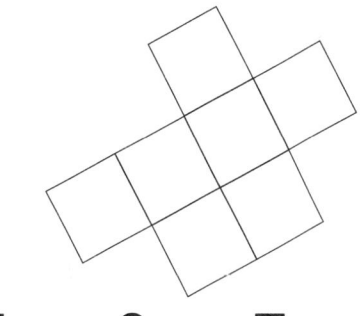

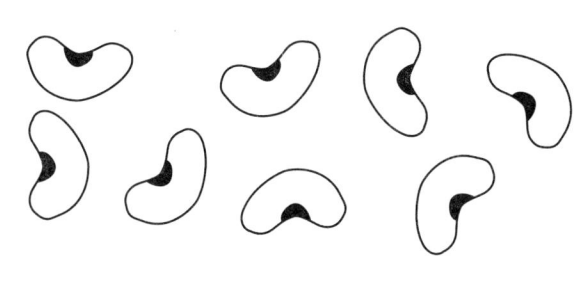

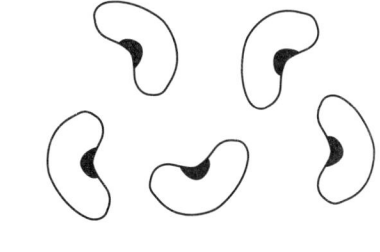

5 6 7 8	5 6 7 8
5 6 7 8	5 6 7 8
5 6 7 8	5 6 7 8

Name _____

Write the number.

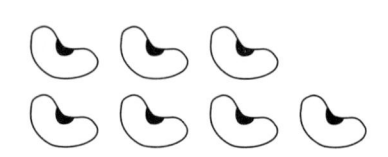

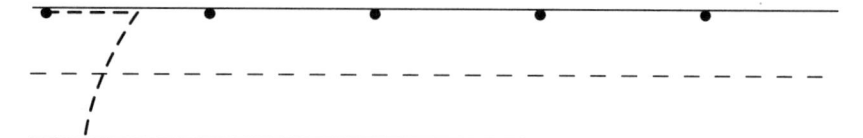

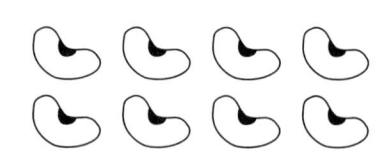

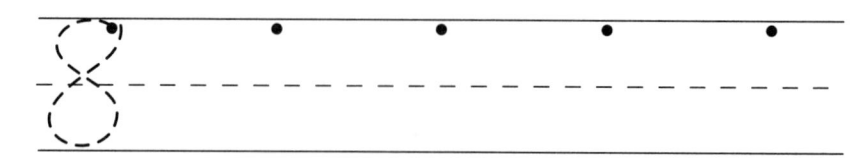

How many? Write the number.

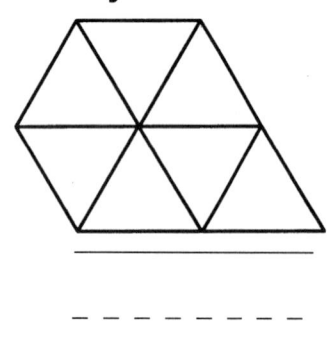

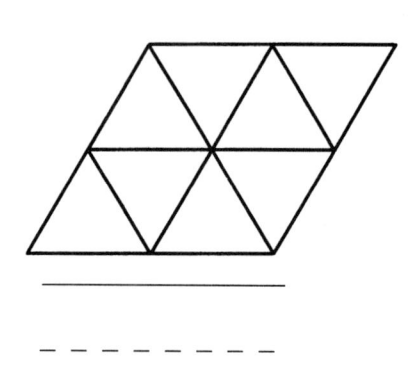

How many? Use 2 🖍 to color the ⊂s.

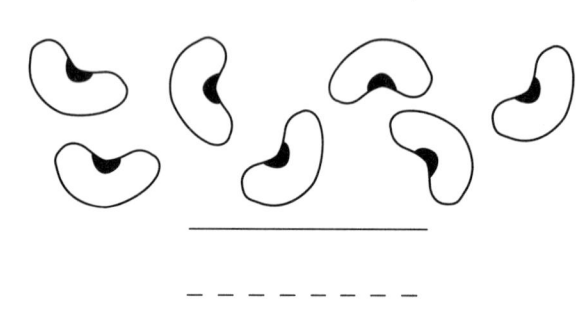

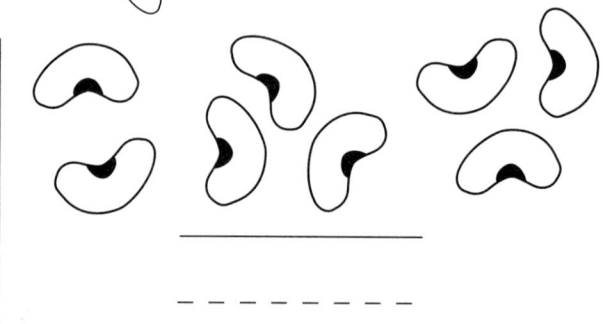

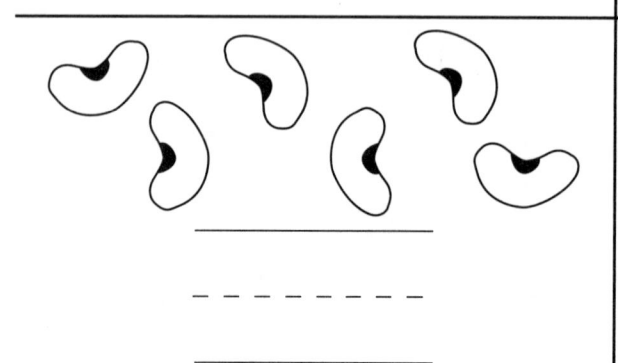

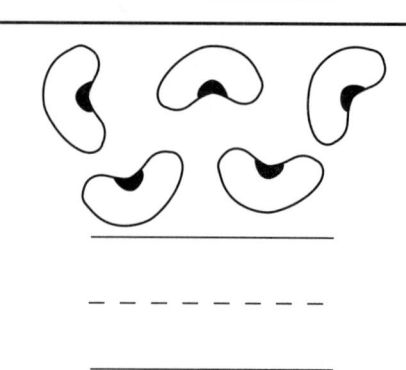

Name _____

Nine and Ten

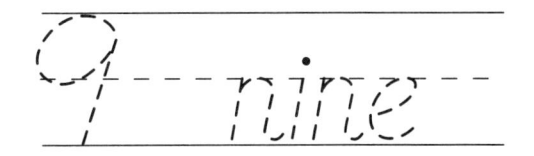

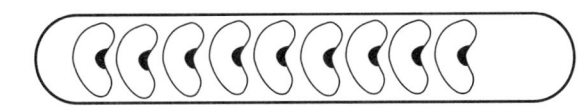

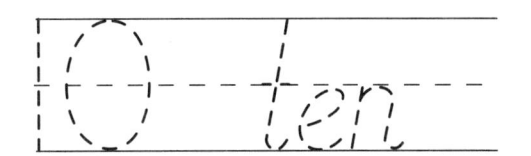

9 nine

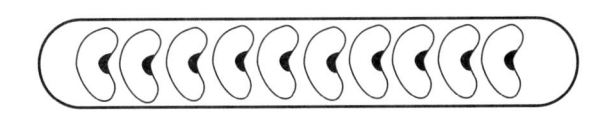

10 ten

How many? Ring the number.

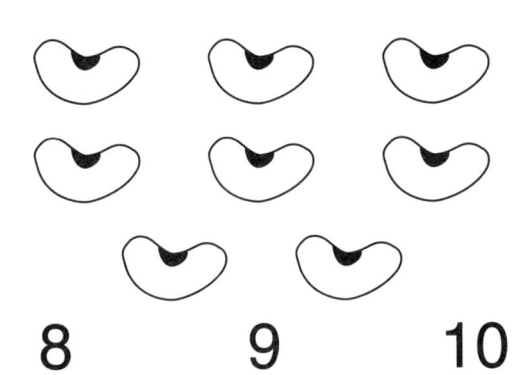

8   9   10

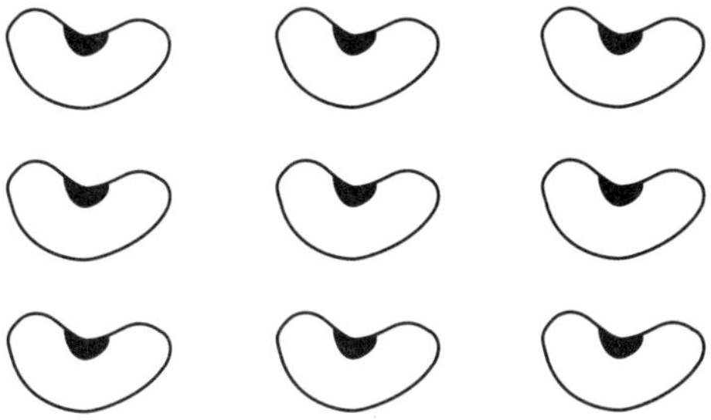

8   9   10

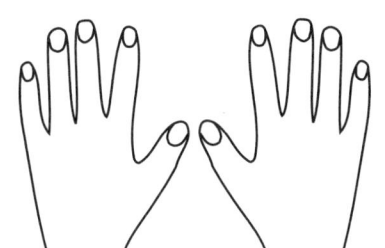

8   9   10

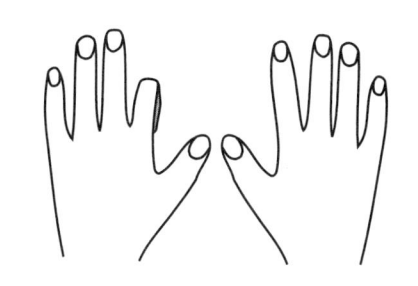

8   9   10

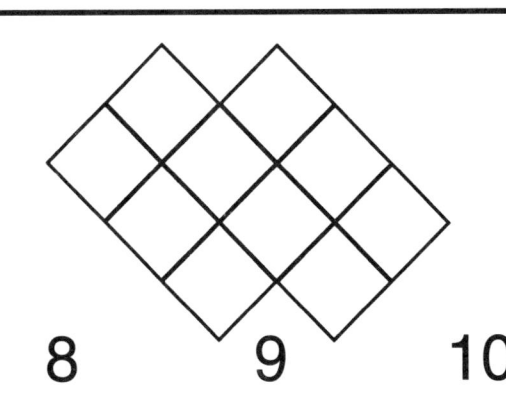

8   9   10

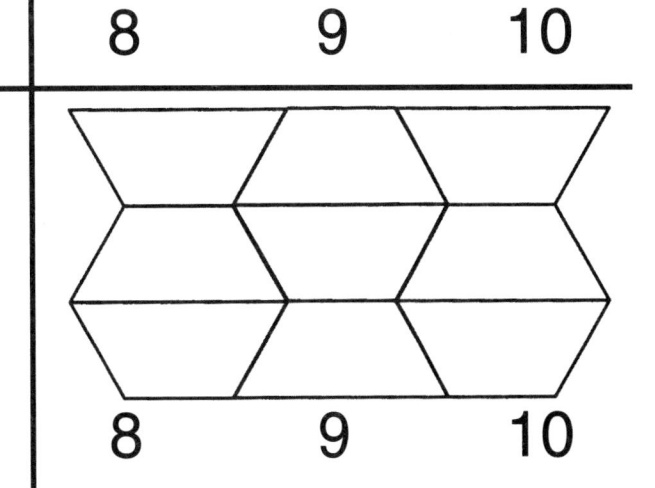

8   9   10

Name _____

Write the number.

How many 🐭 in the lake? Write the number.

How many 🐭 in the ⛺ and in the 🏞 altogether?

46
© Math Teachers Press, Inc., Reproduction by any means is strictly prohibited.

Ordering Numbers from 1 to 10

Name _____

Before, After, Between

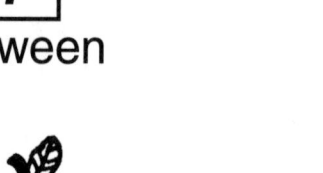

6	7	8
Before	Between	After

What number comes after?	What number comes between?
3 4 ___	5 ___ 7
4 5 6 7	3 4 6 8
What number comes before?	What number comes between?
___ 9 10	4 ___ 6
6 7 8 9	3 5 7 8
What number comes after?	What number comes before?
8 9 ___	___ 6 7
6 7 9 10	4 5 8 9

Name _____

Ordinal Numbers

first second third fourth fifth sixth seventh eighth ninth tenth

Connect the word that matches the number.

2	fifth
5	first
3	sixth
9	second
6	ninth
1	third
8	tenth
7	fourth
10	eighth
4	seventh

Name _____

Bears on Parade

The shaded 🐻 is _____?

first second third
○   ○   ○

The shaded 🐻 is _____?

first second third
○   ○   ○

The shaded 🚗 is _____?

third fourth fifth
○   ○   ○

The shaded 🚗 is _____?

third fourth fifth
○   ○   ○

The shaded 🐴 is _____?

fifth sixth seventh
○   ○   ○

The shaded 🐴 is _____?

fifth sixth seventh
○   ○   ○

The shaded 🦆 is _____?

eighth ninth tenth
○   ○   ○

The shaded 🦆 is _____?

eighth ninth tenth
○   ○   ○

50
© Math Teachers Press, Inc., Reproduction by any means is strictly prohibited.

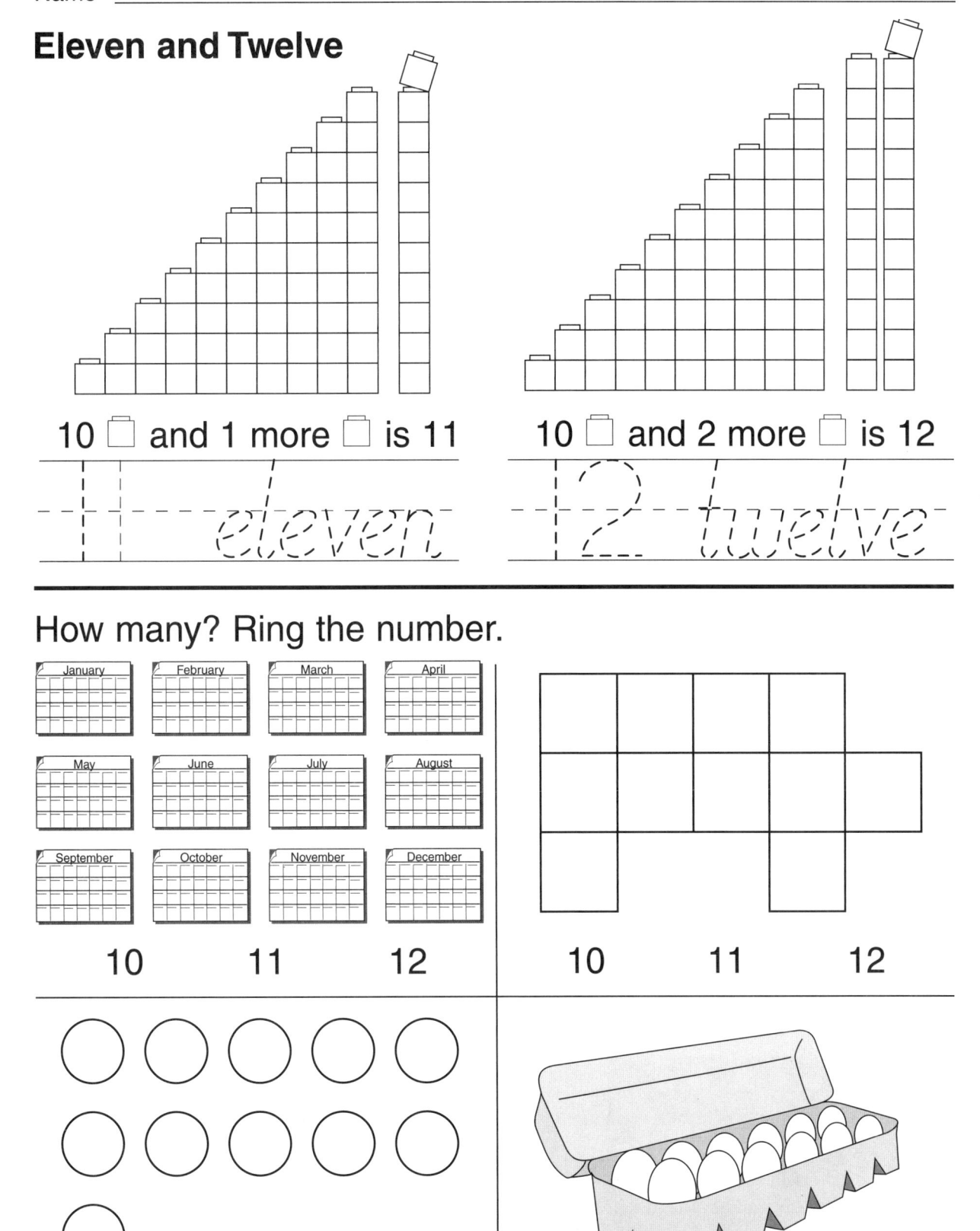

Name _____

Eleven and Twelve

10 ▢ and 1 more ▢ is 11

11 eleven

10 ▢ and 2 more ▢ is 12

12 twelve

How many? Ring the number.

10   11   12

10   11   12

10   11   12

10   11   12

Name _____

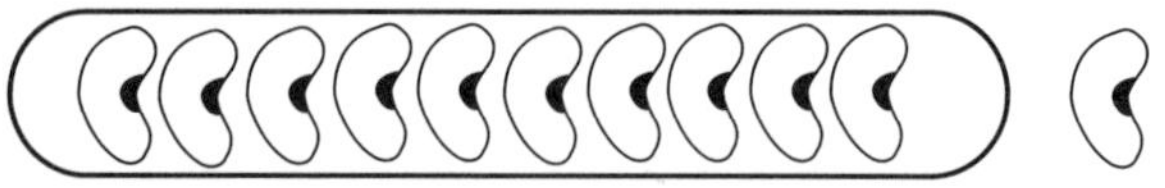

11 eleven

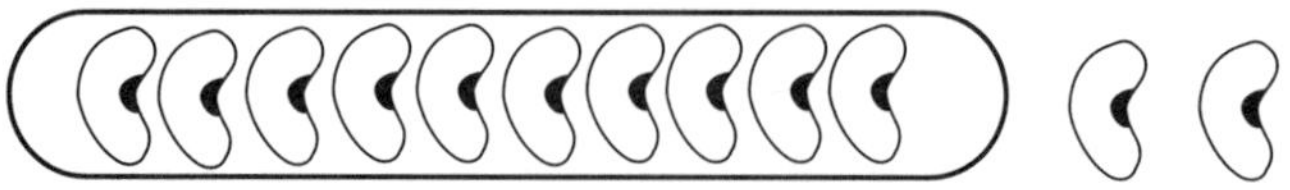

12 twelve

How many? Write the number.

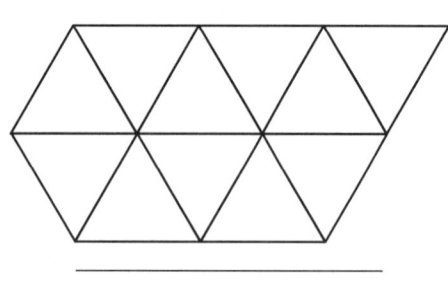

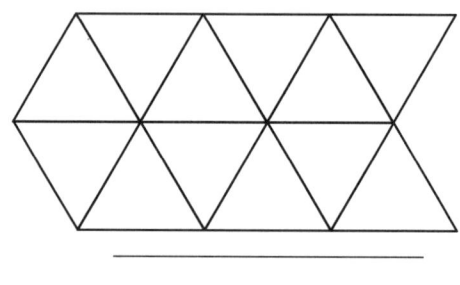

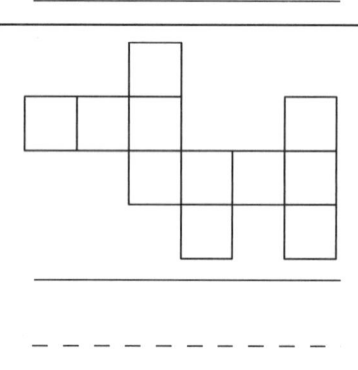

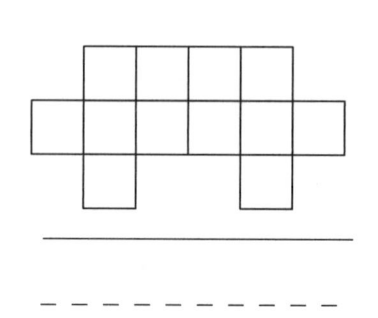

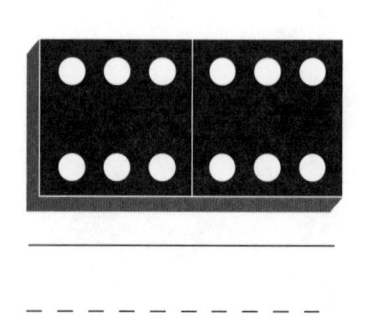

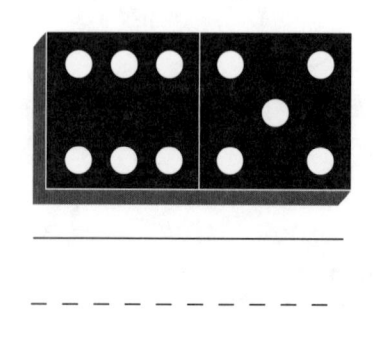

Name _____

More

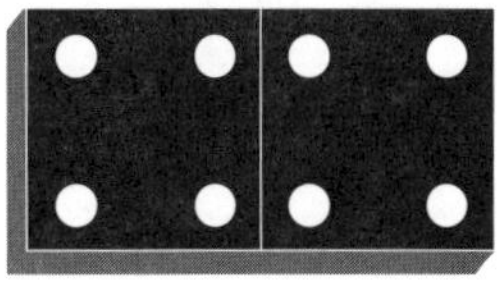

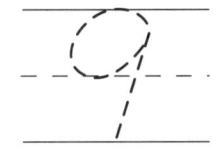

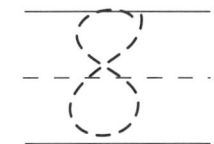

___9___ is more than ___8___

Which is more?

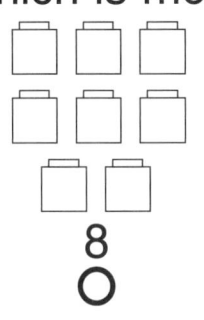

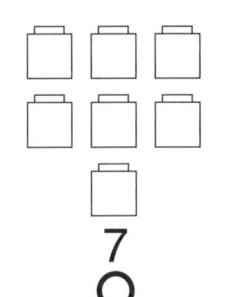

8 7
○ ○

Which is more?

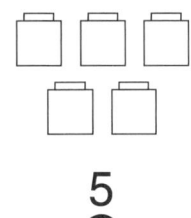

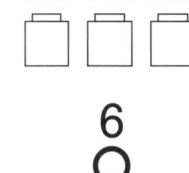

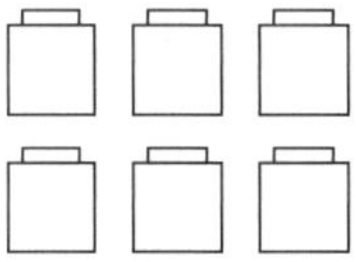

5 6
○ ○

How many? Ring more.

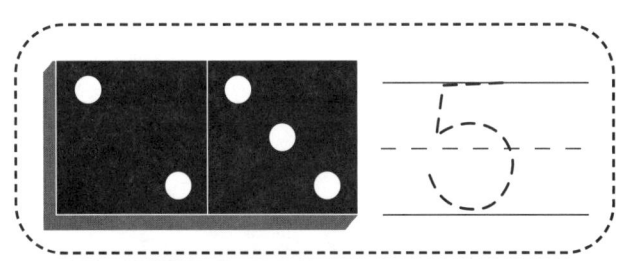

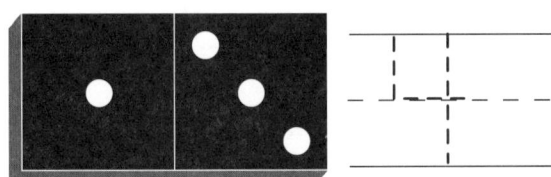

4   ⑤

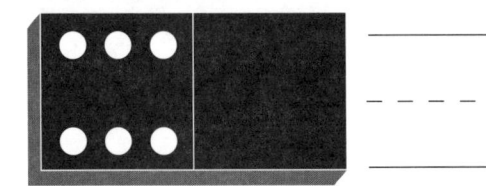

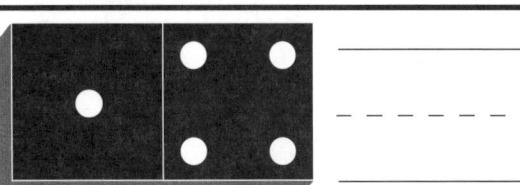

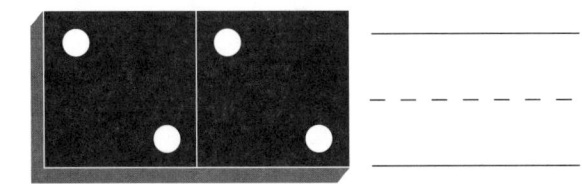

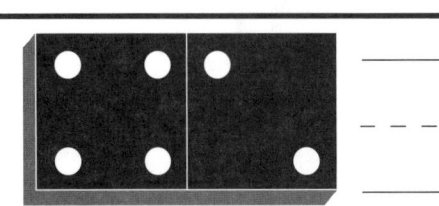

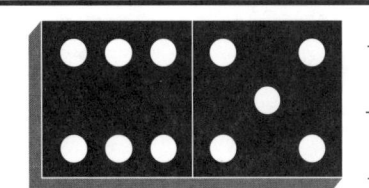

Name _____

Less

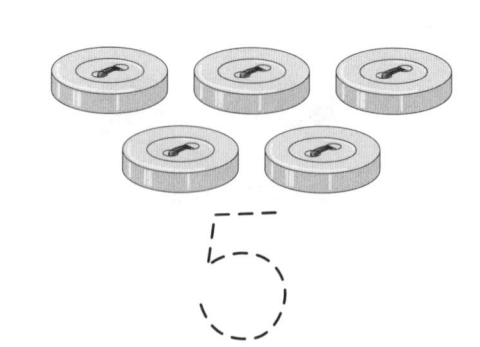

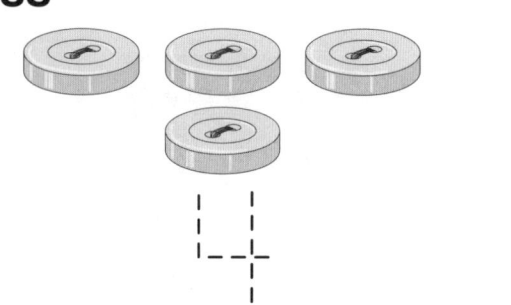

4   5

How many? Ring less.

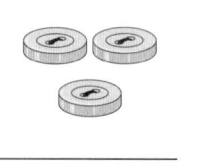

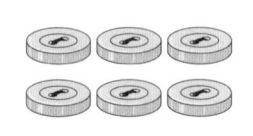

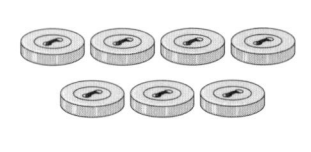

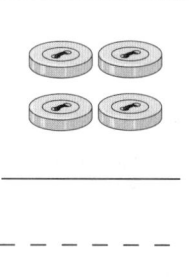

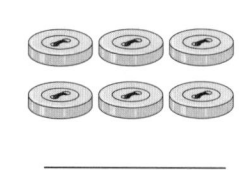

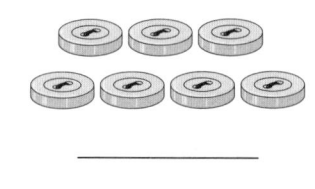

___ ___   ___ ___
--- ---   --- ---
___ ___   ___ ___

___ ___   ___ ___
--- ---   --- ---
___ ___   ___ ___

How many? Ring less.

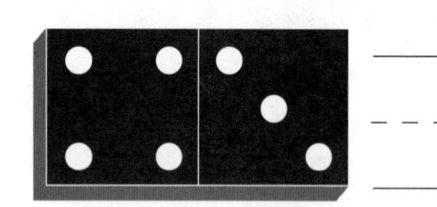

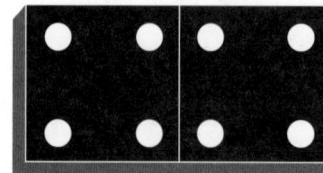

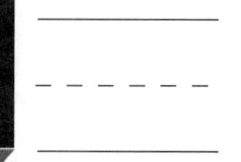

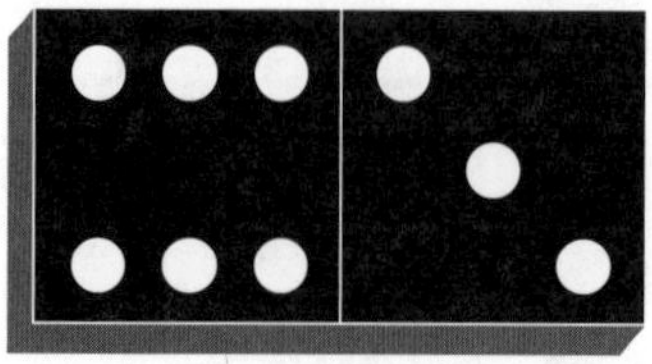

___   ___
---   ---
___   ___

___   ___
---   ---
___   ___

Name _____

Pennies

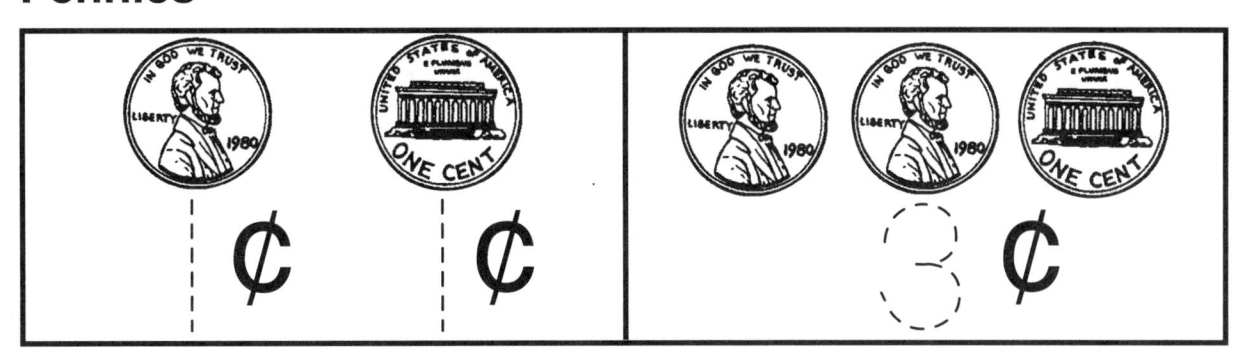

| 🪙 🪙 2¢ | 🪙 🪙 🪙 3¢ |

Write how much money.

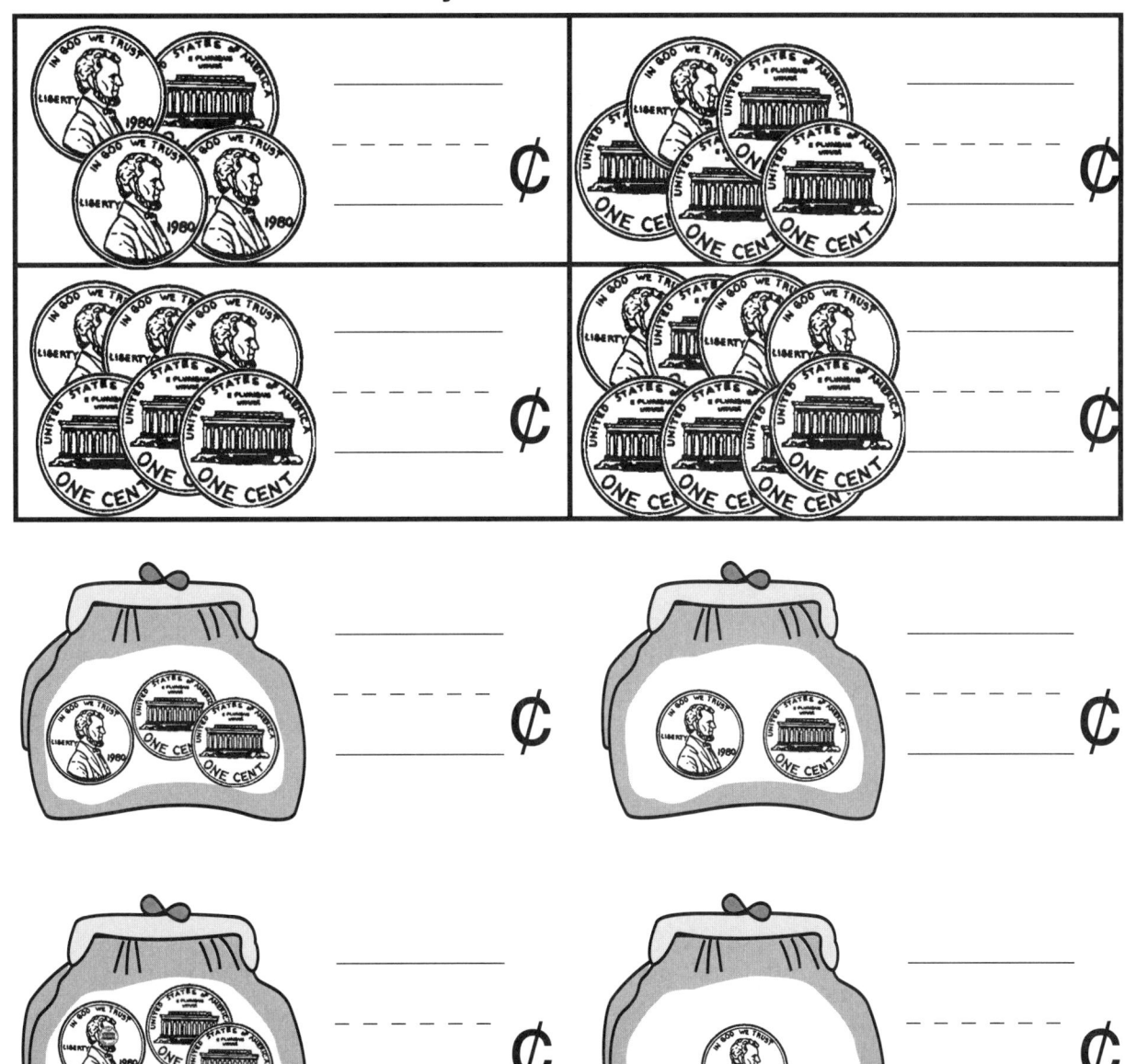

| _____ ¢ | _____ ¢ |
| _____ ¢ | _____ ¢ |

| _____ ¢ | _____ ¢ |
| _____ ¢ | _____ ¢ |

Name _____

car	7¢
airplane	5¢
boat	3¢
school bus	6¢
train	8¢

Ring enough pennies.

You have 7¢. Could you buy the train?   ○ Yes   ○ No

You have 5¢. Could you buy the school bus?   ○ Yes   ○ No

You have 6¢. Could you buy the airplane?   ○ Yes   ○ No

You have 5¢. Which could you buy?

○   ○   ○

Name _____

🖍 **crayon** the matching key.

Name _____

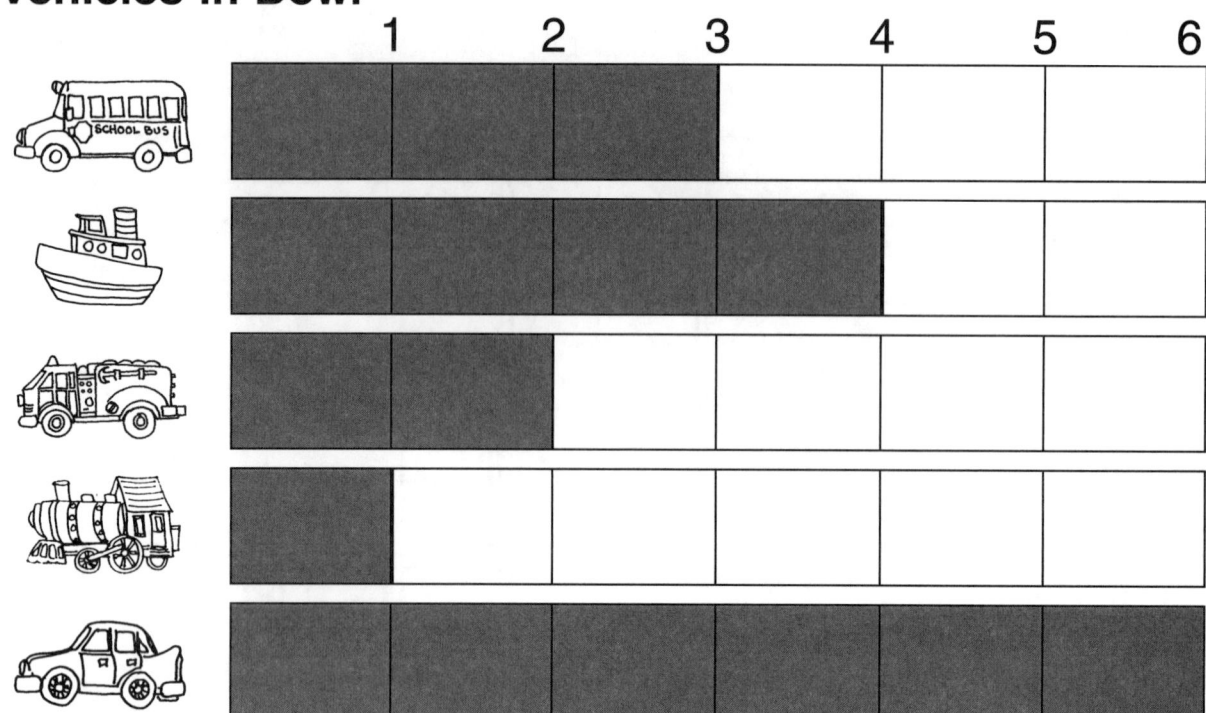

Vehicles in Bowl

Look at the graph.

How many 🚂 ? _____

How many 🚗 ? _____

How many 🚒 ? _____

How many 🚌 ? _____

How many ⛵ ? _____

Which is more? 🚗 ⛵
           ○   ○

Which is more? 🚂 🚒
           ○   ○

Which is less? 🚌 ⛵
           ○   ○

Name _____

Chapter 2 Review

1. How many?

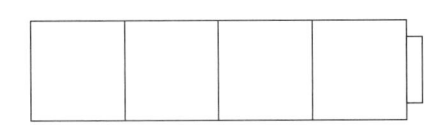

    1   2   3   4

2. How many?

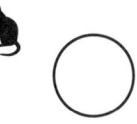

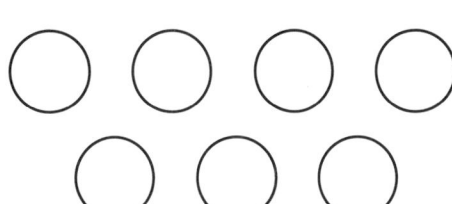

    5   6   7   8

3. How many in all?

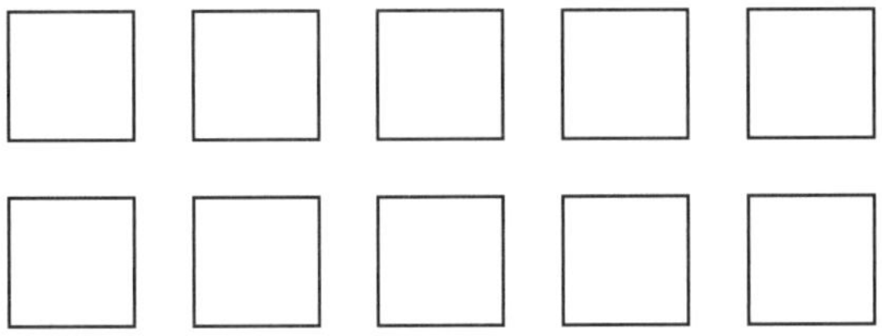

    9   10   11   12

4. How many in all?

    9   10   11   12

5. Write how many.

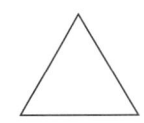

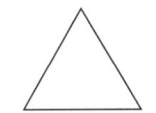

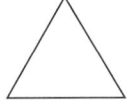

_____

6. Write how many.

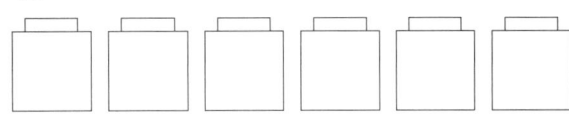

_____

7. Write how many.

_____

8. Write how many in all.

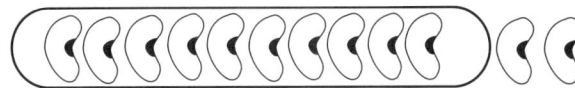

_____

Name _____

Chapter 2 Review

9. What number comes after?

7 8 ____

6   7   8   9

10. What number comes in the middle?

5 ____ 7

4   5   6   7

11. The shaded 🐻 is ____?

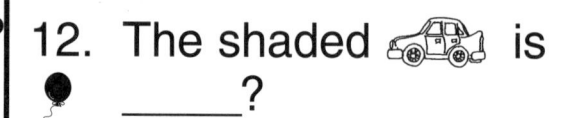

first   second   third

12. The shaded 🚗 is ____?

fourth   fifth   sixth

13. Ring more.

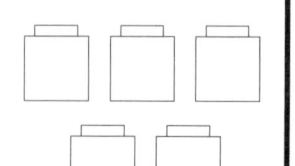

6   5

14. Ring less.

4   3

15. Write how much money.

_____ ¢

16. Ring enough pennies.

60
© Math Teachers Press, Inc., Reproduction by any means is strictly prohibited.

3
Sums to 10

Name _____

Write how many.

Name _____

One More

3   4   5

1. _____   _____   _____

2. _____   _____   _____

3. _____   _____   _____

63
© Math Teachers Press, Inc. Reproduction by any means is strictly prohibited.

Name _____

Two More, Three More
How many? Tell the story.

3   2   5

1.

_____   _____   _____   _____

2.

_____   _____   _____   _____

3.

_____   _____   _____   _____

64
© Math Teachers Press, Inc. Reproduction by any means is strictly prohibited.

Name _____

Plus
How many in all?

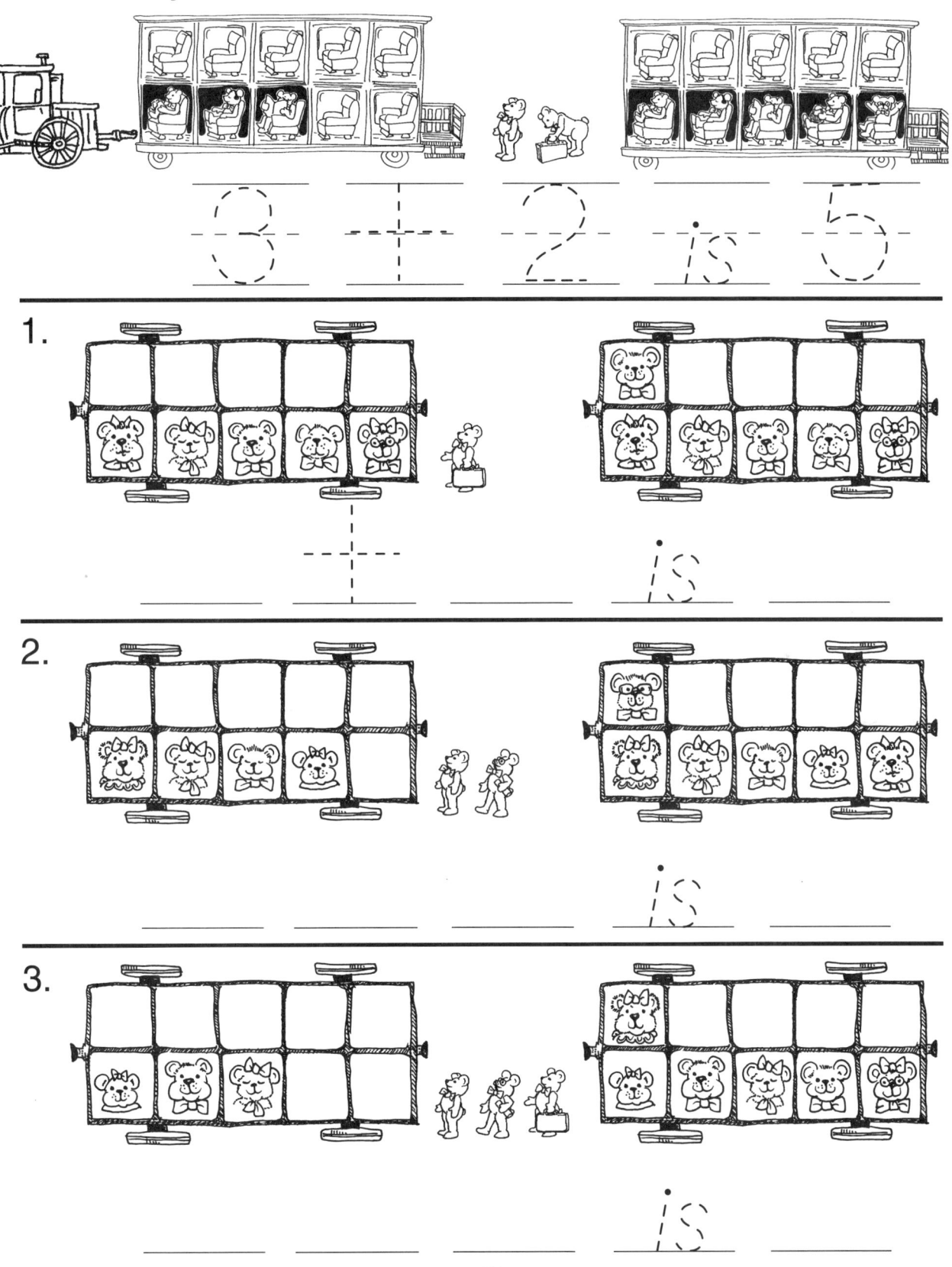

3 + 2 is 5

1. ___ + ___ is ___

2. ___ ___ ___ is ___

3. ___ ___ ___ is ___

65
© Math Teachers Press, Inc. Reproduction by any means is strictly prohibited.

Name _____

Equals, Is the Same As, Sum

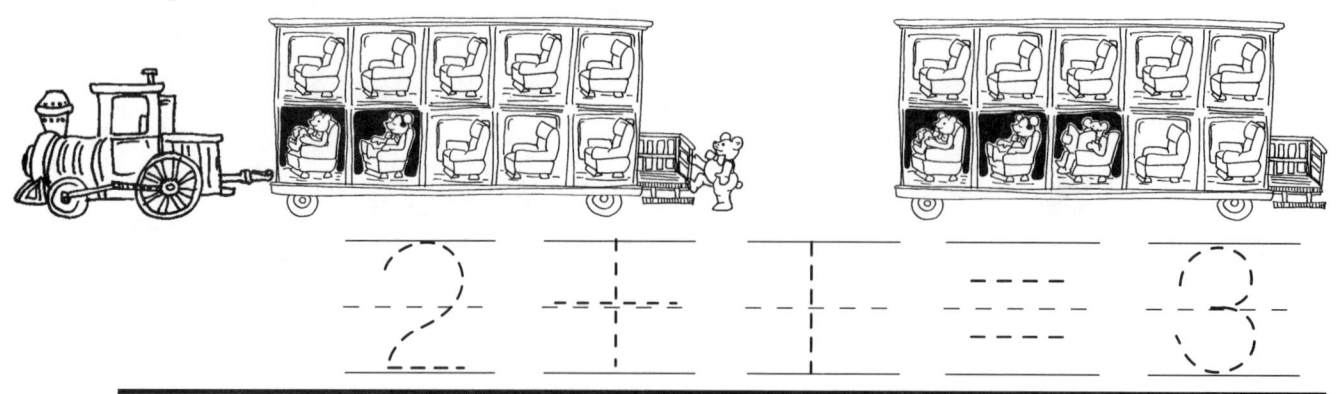

2 + 1 = 3

Tell a story. Draw a picture. Write the sentence.

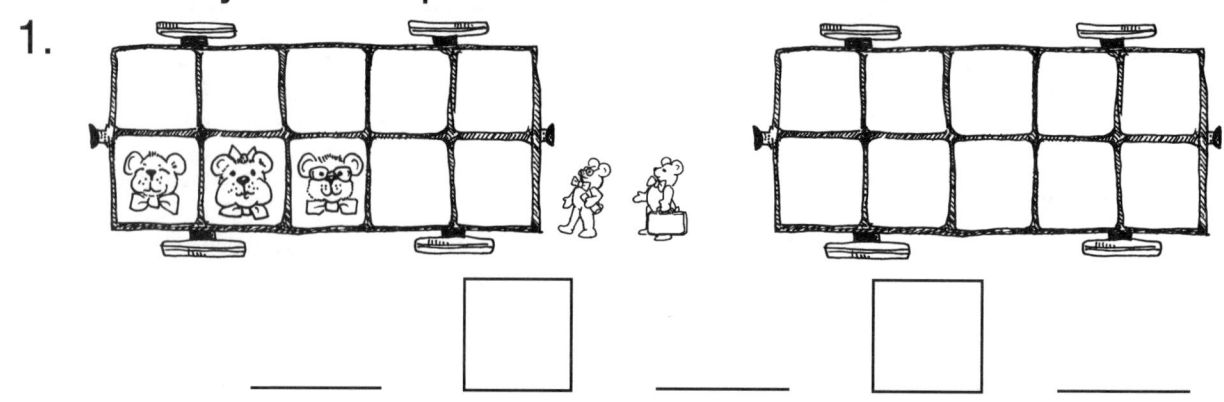

1.

2.

3.

66
© Math Teachers Press, Inc. Reproduction by any means is strictly prohibited.

Name _____

Dinosaur Count

2 + 2 = _____

Tell the story. Write the number. Then add.

1.

☐ ___ ☐ ___

2.

☐ ___ ☐ ___

3.

☐ ___ ☐ ___

4.

☐ ___ ☐ ___

5.   3 + 3 = _____   5 + 1 = _____   4 + 1 = _____

6.   2 + 2 = _____   2 + 4 = _____   3 + 1 = _____

Name _____

Stegosaurus

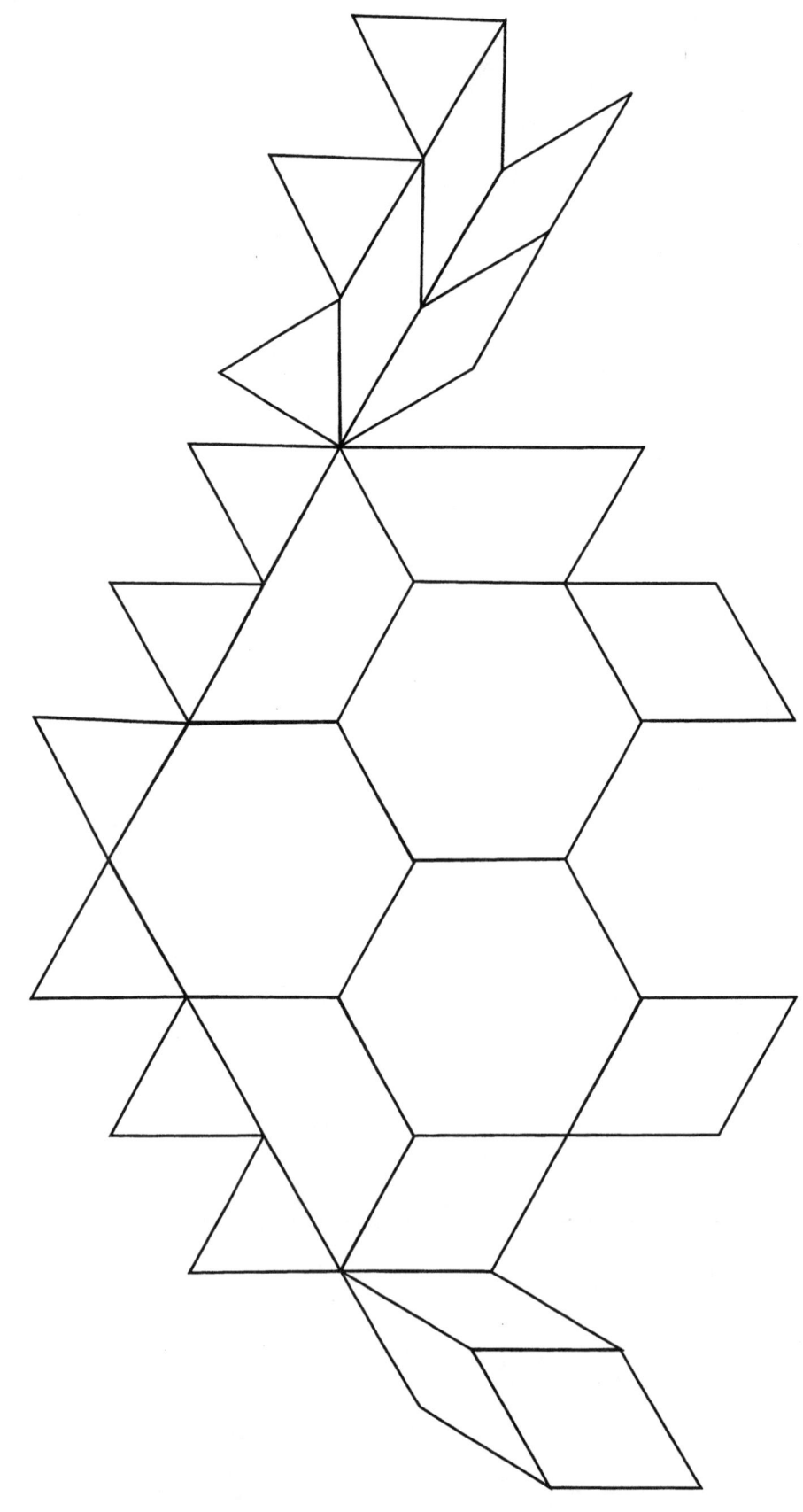

Cover with pattern blocks. Make a graph. How many?

Name _____

Waiting on the Train: Adding with Zero

____ + ____ = ____

Write a number sentence.

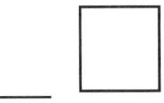

1.

2.

3.

4.

5. 3 + 1 = _____   2 + 0 = _____   0 + 5 = _____

6. 3 + 0 = _____   3 + 3 = _____   4 + 2 = _____

69

Name _____

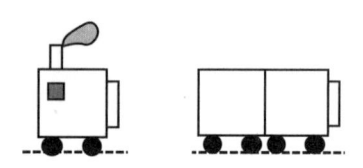

Zero is the name for none or no cars.

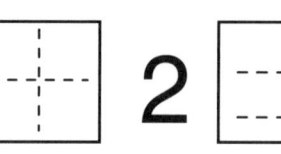

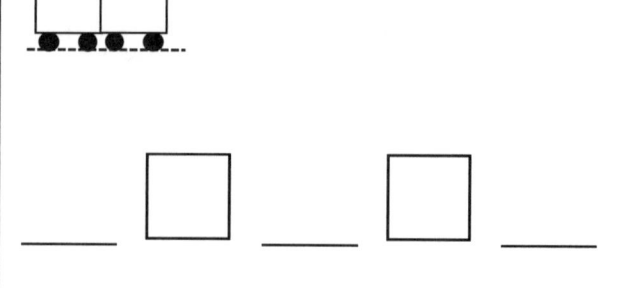

1 ☐ 2 ☐ 3   ___ ☐ ___ ☐ ___

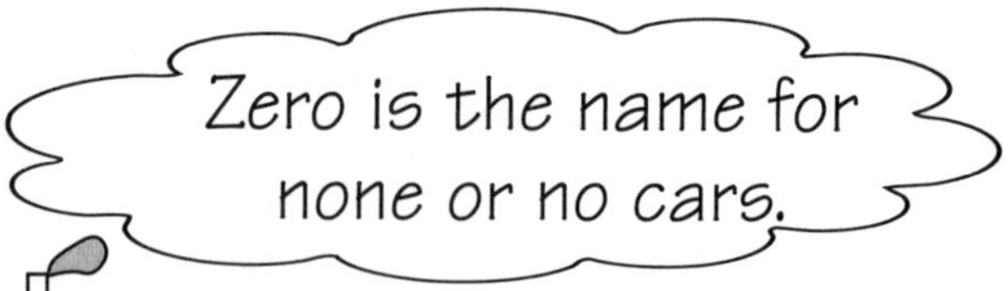

Write a number sentence.

1.

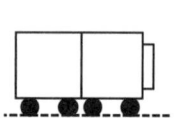

___ ☐ ___ ☐ ___

2.

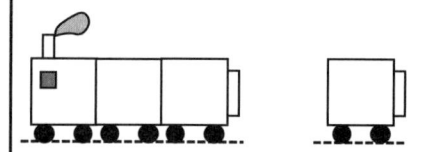

___ ☐ ___ ☐ ___

3.

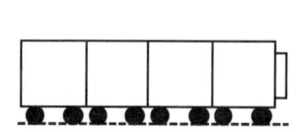

___ ☐ ___ ☐ ___

4.

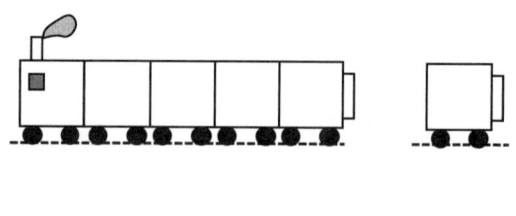

___ ☐ ___ ☐ ___

5.   4 + 2 = _____   5 + 1 = _____   0 + 0 = _____

6.   2 + 3 = _____   4 + 0 = _____   0 + 5 = _____

Name _____

What Happens When?

1. 3 apples
   2 apples
   How many in all?

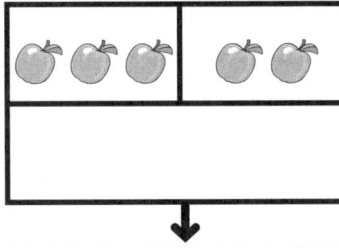

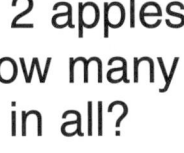

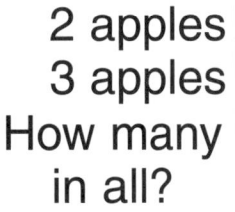

   Draw a picture.

   3 + 2 = _____

   2 apples
   3 apples
   How many in all?

   Draw a picture.

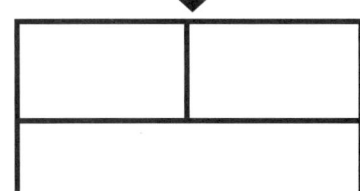

   2 + 3 = _____

   The sums are the _____.

2. 4 nuts
   2 nuts

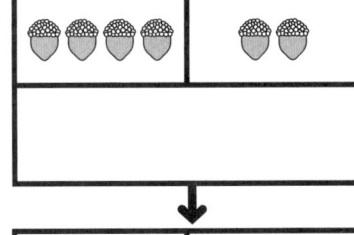

   How many in all?
   Draw all.

   4 + 2 = _____

   2 nuts
   4 nuts

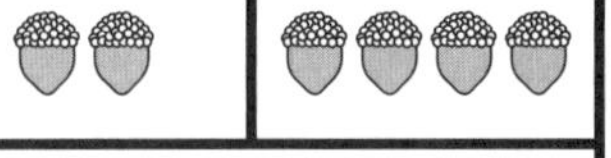

   How many in all?
   Draw all.

   2 + 4 = _____

   The sums are the _____.

3. Match the sums.

   3 + 0 = _____   4 + 1 = _____

   1 + 4 = _____   1 + 5 = _____

   0 + 2 = _____   0 + 3 = _____

   5 + 1 = _____   2 + 0 = _____

Name _____

It is easier to start with the bigger number: 5 + 1 = 6

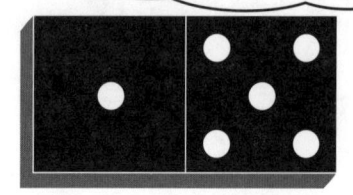

and

1 + 5 = 6   5 + 1 = 6

Draw a picture of the related domino.

1.   =

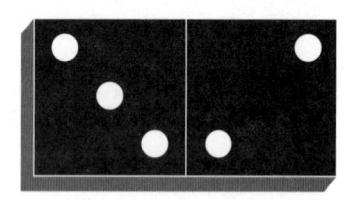

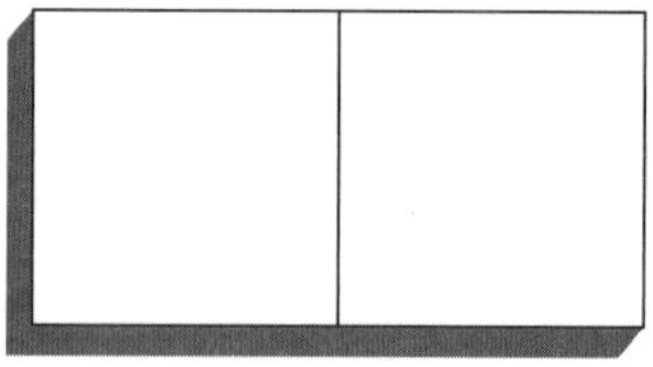

   3 + 2   =   ____ + ____

2.   =

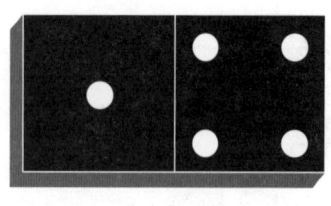

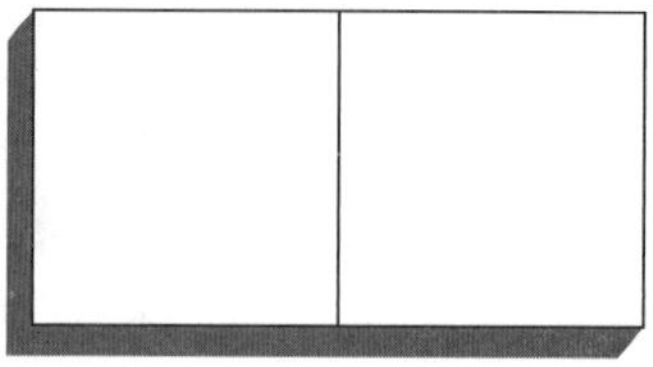

   1 + 4   =   ____ + ____

Ring the greater number. Add by counting on.

3. 4 + 2 = ____   4, 5, 6   2 + 4 = ____   4, 5, 6

4. 1 + 3 = ____   3, 4   3 + 1 = ____   3, 4

5. 2 + 3 = ____   3 + 2 = ____

Name _____

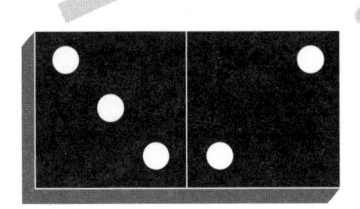

Turn the domino.

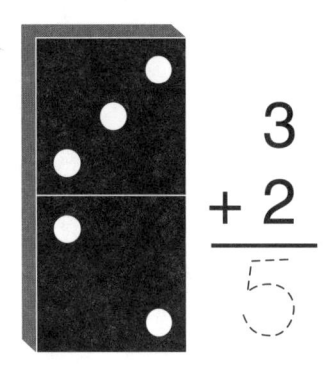

3 + 2 = 5

3
+ 2

5

1. ___ + ___ = ___   + ___

2. ___ + ___ = ___   + ___

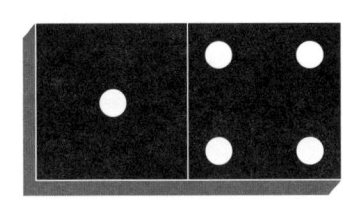

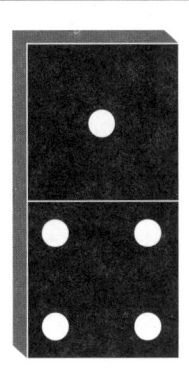

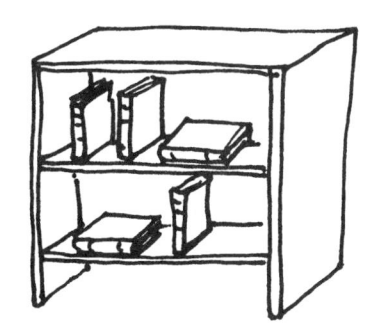

Add.

3. ○   1   △△△   3   ☐☐☐
   ○○○○○ + 5   △△△ + 3   ☐☐ + ___
              ---   ---
               6

4. △   1   ☐☐   ○○○
   △△   + 2   ☐☐ + ___   + ___

Name _____

> 3 and 1 more is 4.

$$\begin{array}{r} 3 \\ +\ 1 \\ \hline 4 \end{array}$$

Draw a picture of each number.
Add. Draw a picture of the answer.

1.
$$\begin{array}{r} 2 \\ +\ 1 \\ \hline 3 \end{array}$$
$$\begin{array}{r} 4 \\ +\ 0 \\ \hline \end{array}$$
$$\begin{array}{r} 3 \\ +\ 1 \\ \hline \end{array}$$

2.
$$\begin{array}{r} 5 \\ +\ 1 \\ \hline \end{array}$$
$$\begin{array}{r} 4 \\ +\ 2 \\ \hline \end{array}$$
$$\begin{array}{r} 0 \\ +\ 3 \\ \hline \end{array}$$

3.
$$\begin{array}{r} 3 \\ +\ 2 \\ \hline \end{array}$$
$$\begin{array}{r} 2 \\ +\ 2 \\ \hline \end{array}$$
$$\begin{array}{r} 4 \\ +\ 1 \\ \hline \end{array}$$

Name _____

Part Plus Part Equals Whole

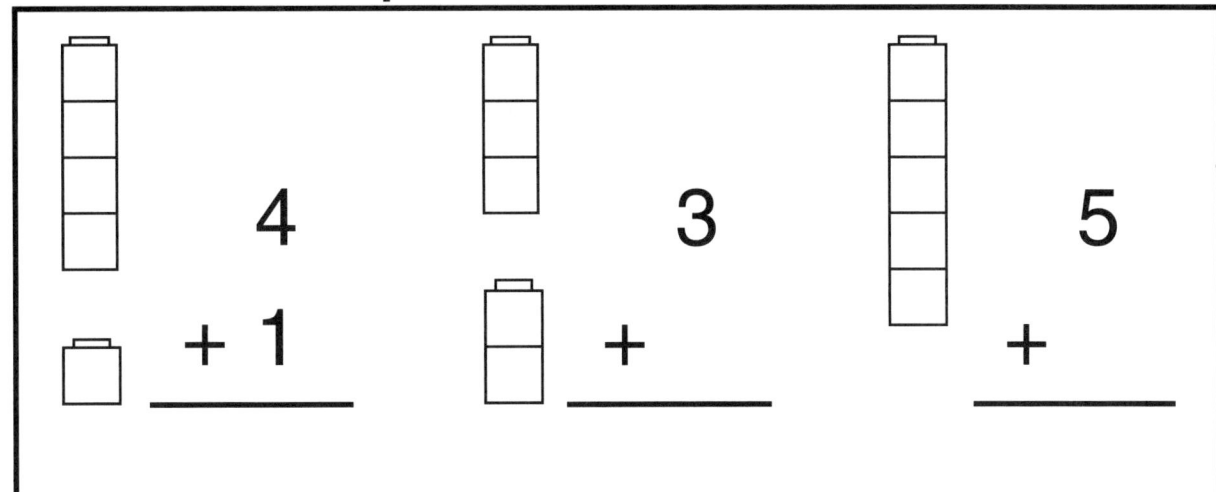

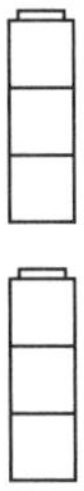

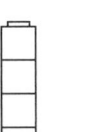

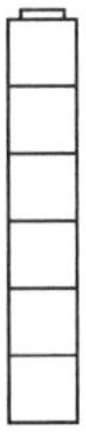

$\quad 4\quad\quad\quad 3\quad\quad\quad 5$
$\underline{+\ 1}\quad\ \ \underline{+}\quad\ \ \underline{+}$

1.
$\quad\quad 2\quad\quad\quad\quad 1\quad\quad\quad\quad 2$
$\underline{+\ 2}\quad\quad\ \underline{+\ 3}\quad\quad\ \underline{+\ 0}$

2.
$\quad\quad 1\quad\quad\quad\quad 2\quad\quad\quad\quad 3$
$\underline{+\ 1}\quad\quad\ \underline{+\ 4}\quad\quad\ \underline{+\ 1}$

3.
$\quad\quad 3\quad\quad\quad\quad 5\quad\quad\quad\quad 0$
$\underline{+\ 3}\quad\quad\ \underline{+\ 1}\quad\quad\ \underline{+\ 6}$

Name _____

Sums to 6
How many do you know?

2	1	3	0	2	3
+1	+4	+0	+6	+0	+2

1	0	1	0	4	3
+1	+5	+5	+1	+0	+3

3	2	0	4	0	6
+1	+4	+0	+1	+3	+0

2	2	5	0	5	1
+3	+2	+1	+2	+0	+3

0	1	4	1
+4	+0	+2	+2

No. Correct (28)

Name _____

Number Line Addition

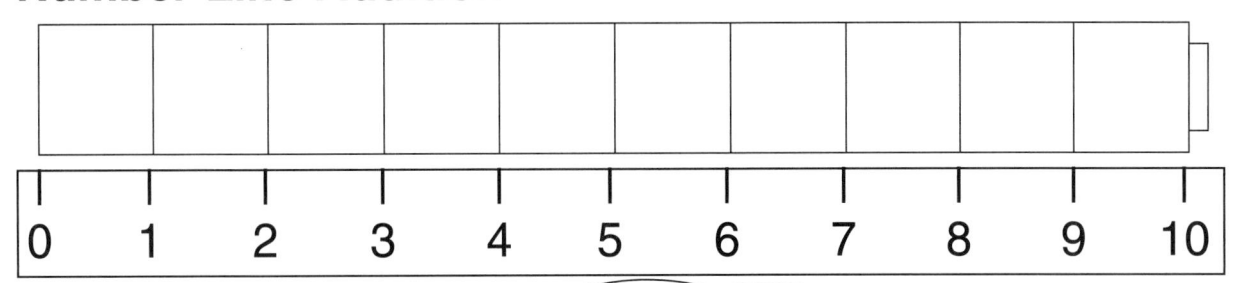

```
|   |   |   |   |   |   |   |   |   |   |
0   1   2   3   4   5   6   7   8   9   10
```

6 + 3 = ___   *Start at 6. Count up 3 steps. 6, 7, 8, 9*   6 + 3 = 9

Use the number line to count up 1, 2 or 3.

1.   6   *6, 7*   7   *7, 8, 9*   9   *9, 10*
    +1   +2   +1

2.   8   6   7
    +1   +2   +3

3.   6   7   8
    +3   +1   +2

4.   3   *Start with 6. Count up 3 steps.*   2   1
    +6   +6   +9

5.   2   2   3
    +7   +8   +7

Name _____

Problem Solving
Make a table and find the pattern.

Finish each table.

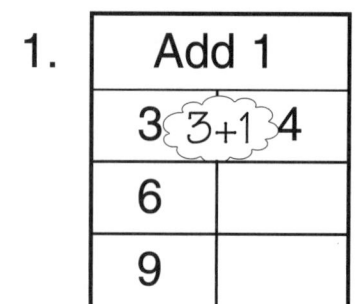

1.
Add 1	
3 (3+1)	4
6	
9	

2.
Add 2	
6	
8	
5	

3.
Add 3	
6	
3	
7	

4.
Add 0	
5	
8	
6	

5.
Add 3	
2	
4	
5	

6.
Add 2	
4	
7	
2	

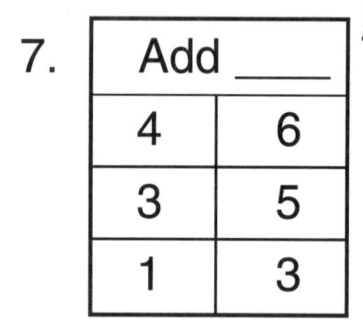

Find the pattern.

7.
Add ___	
4	6
3	5
1	3

8.
Add ___	
6	7
3	4
8	9

9.
Add ___	
2	5
6	9
4	7

10.
___	___
5	5
4	4
7	7

11.
___	___
5	7
2	4
6	8

12.
___	___
5	8
3	6
7	10

Name _____

Doubles

The double of 3 is 6.
3 + 3 = 6

1.   1   3   2   0   4   5
     +1   +3   +2   +0   +4   +5

2. 3 + 3 = _____   0 + 0 = _____   1 + 1 = _____

3. 2 + 2 = _____   5 + 5 = _____   4 + 4 = _____

Math Machine

4.

Double	
In	Out
4	
3	
5	
2	
1	
0	

5.

Add 3	
In	Out
7	
6	
5	
4	
3	
2	

6.

Add 0	
In	Out
9	
4	
6	
0	
8	
5	

© Math Teachers Press, Inc. Reproduction by any means is strictly prohibited.

Name _____

Adding a Number and its Neighbor

3 + 4

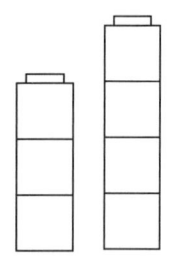

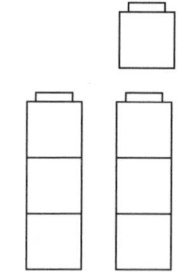

Double 3 and add 1 more.

3 + 4 = double 3 + 1
       = 6 + 1
       = 7

1. 4 + 5

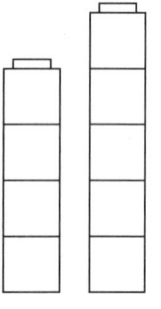

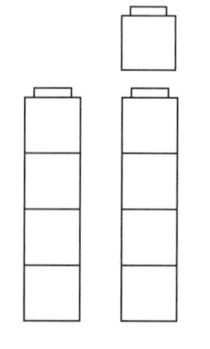

Double 4 and add 1 more.

4 + 5 = double 4 + 1
       = 8 + 1

4 + 5 = _____

2.   2   3   5   6   3   4
    +2   +2   +0   +2   +3   +3

3.   3   8   5   3   3   4
    +7   +2   +5   +6   +4   +4

4. Can you use doubles to find 4 + 6? Explain.

Name _____

Using a Calculator: Counting On to Add

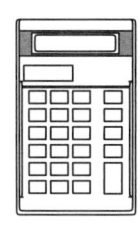

1. Use your calculator to count 1 more.

   [on/c] [on/c] [0] [+] [1] [=] _____ [=] _____

2. Count 1 more.

   [on/c] [4] [+] [1] [=] _____ [=] _____

3. Count 2 more.

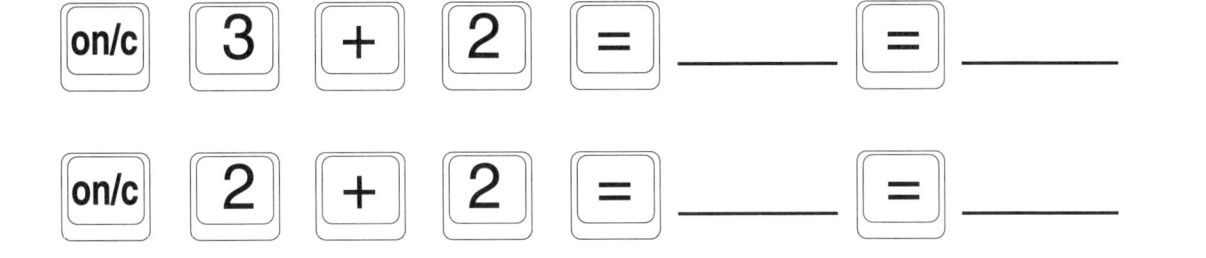

   [on/c] [3] [+] [2] [=] _____ [=] _____

   [on/c] [2] [+] [2] [=] _____ [=] _____

4. Count 3 more.

   [on/c] [3] [+] [3] [=] _____ [=] _____

   [on/c] [2] [+] [3] [=] _____ [=] _____

Name _____

Sums to 10 No. Correct (36) _____

Add

1.   7   3   8   7   2   3
   + 0   + 5   + 1   + 1   + 7   + 4

2.   8   6   2   2   9   2
   + 0   + 1   + 6   + 5   + 0   + 8

3.   1   9   5   5   1   5
   + 6   + 1   + 2   + 5   + 7   + 3

4.   7   0   4   4   4   0
   + 2   + 7   + 6   + 3   + 4   + 8

5.   6   8   4   3   1   3
   + 3   + 2   + 5   + 7   + 8   + 6

6.   0   5   6   6   1   7
   + 9   + 4   + 2   + 4   + 9   + 3

© Math Teachers Press, Inc. Reproduction by any means is strictly prohibited.

Name _____

Complete the addition table for the facts to 10.

+	0	1	2	3	4	5	6	7	8	9
0										
1										
2										
3										
4									
5									
6								
7							
8						
9					

How many facts have sums of...

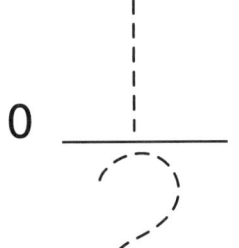

0 _____   3 _____   6 _____   9 _____

1 2   4 _____   7 _____   10 _____

2 _____   5 _____   8 _____

83
© Math Teachers Press, Inc. Reproduction by any means is strictly prohibited.

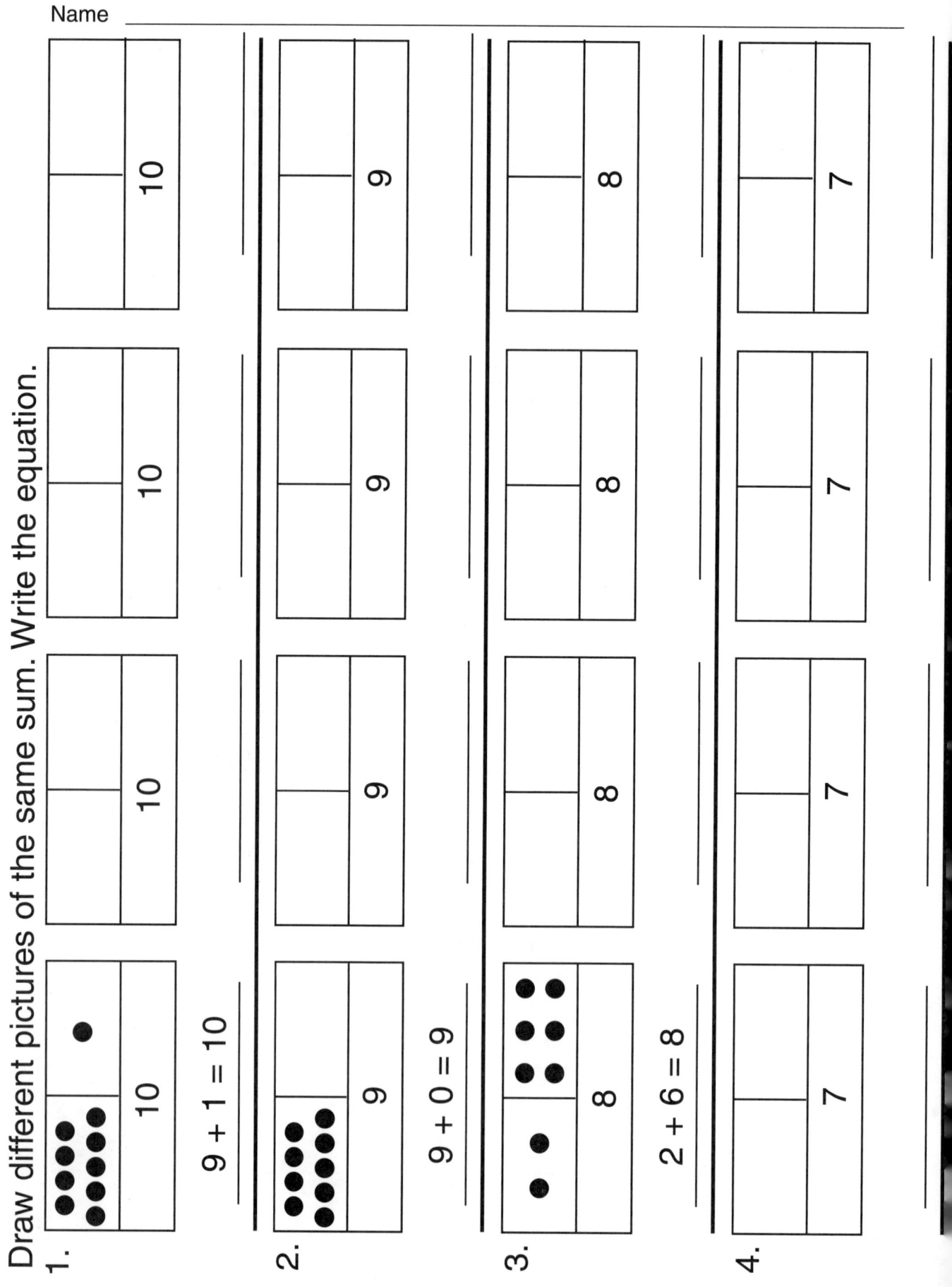

Name _____

Draw different pictures of the same sum. Write the equation.

1. 10 | 10 | 10 | 10

9 + 1 = 10

2. 9 | 9 | 9 | 9

9 + 0 = 9

3. 8 | 8 | 8 | 8

2 + 6 = 8

4. 7 | 7 | 7 | 7

Name _____

Add.
Put a ring around the number of pennies.

1. 2¢ + 3¢ = _____ ¢	
2. 4¢ + 1¢ = _____ ¢	
3. 2¢ + 2¢ = _____ ¢	
4. 3¢ + 3¢ = _____ ¢	
5. 2¢ + 4¢ = _____ ¢	

Name _____

Problem Solving

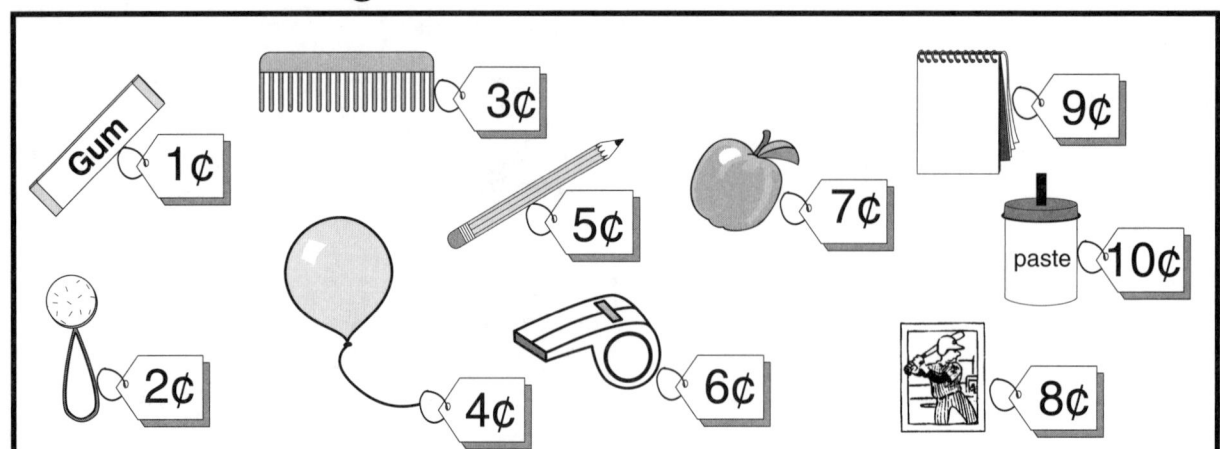

Price list:
- Gum 1¢
- Comb 3¢
- Pencil 5¢
- Apple 7¢
- Notepad 9¢
- Paste 10¢
- Rattle 2¢
- Balloon 4¢
- Whistle 6¢
- Picture 8¢

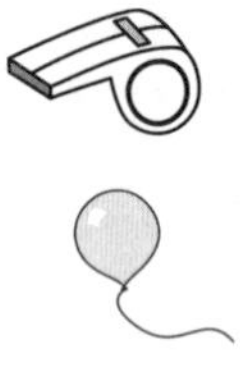

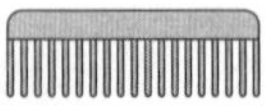

How much in all? Use ¢.

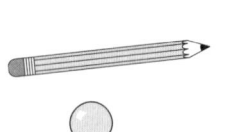

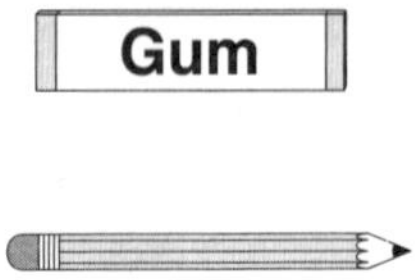

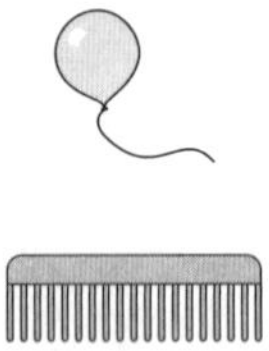

1. Gum ..... 1¢
   Pencil ... + 5¢
   ────────────
            6¢

2. Apple + Rattle = _____

3. Balloon + Comb = _____

4. Picture + Rattle = _____

5. Whistle + Balloon = _____

6. Pencil + Balloon = _____

You have 10¢.

7. Can you buy comb and whistle?
   ○ Yes   ○ No

8. Can you buy notepad and balloon?
   ○ Yes   ○ No

86

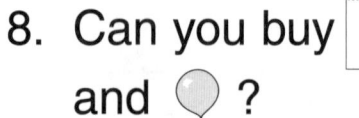

Name _____

How much money is there?

Start at 6...7...8...

6¢

8 ¢

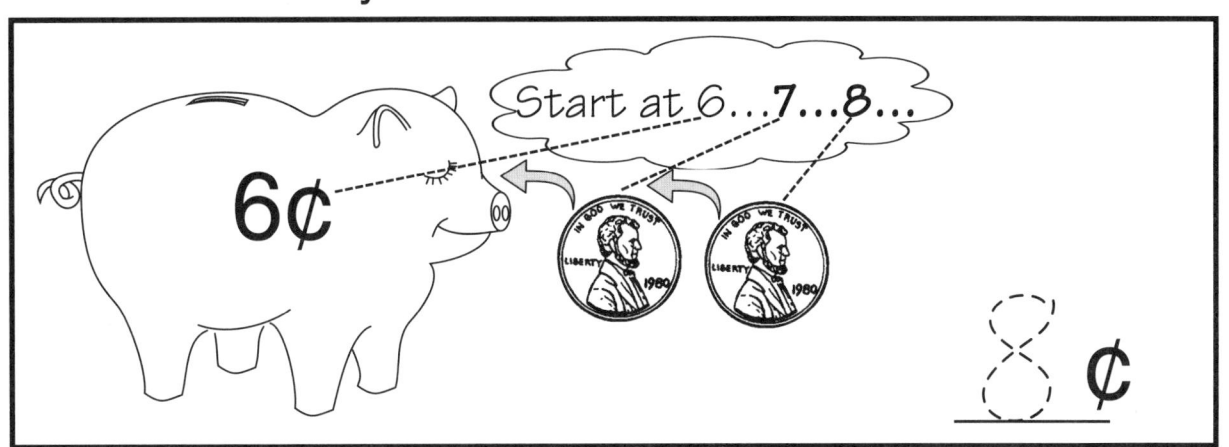

How much money is there? Count on to add.

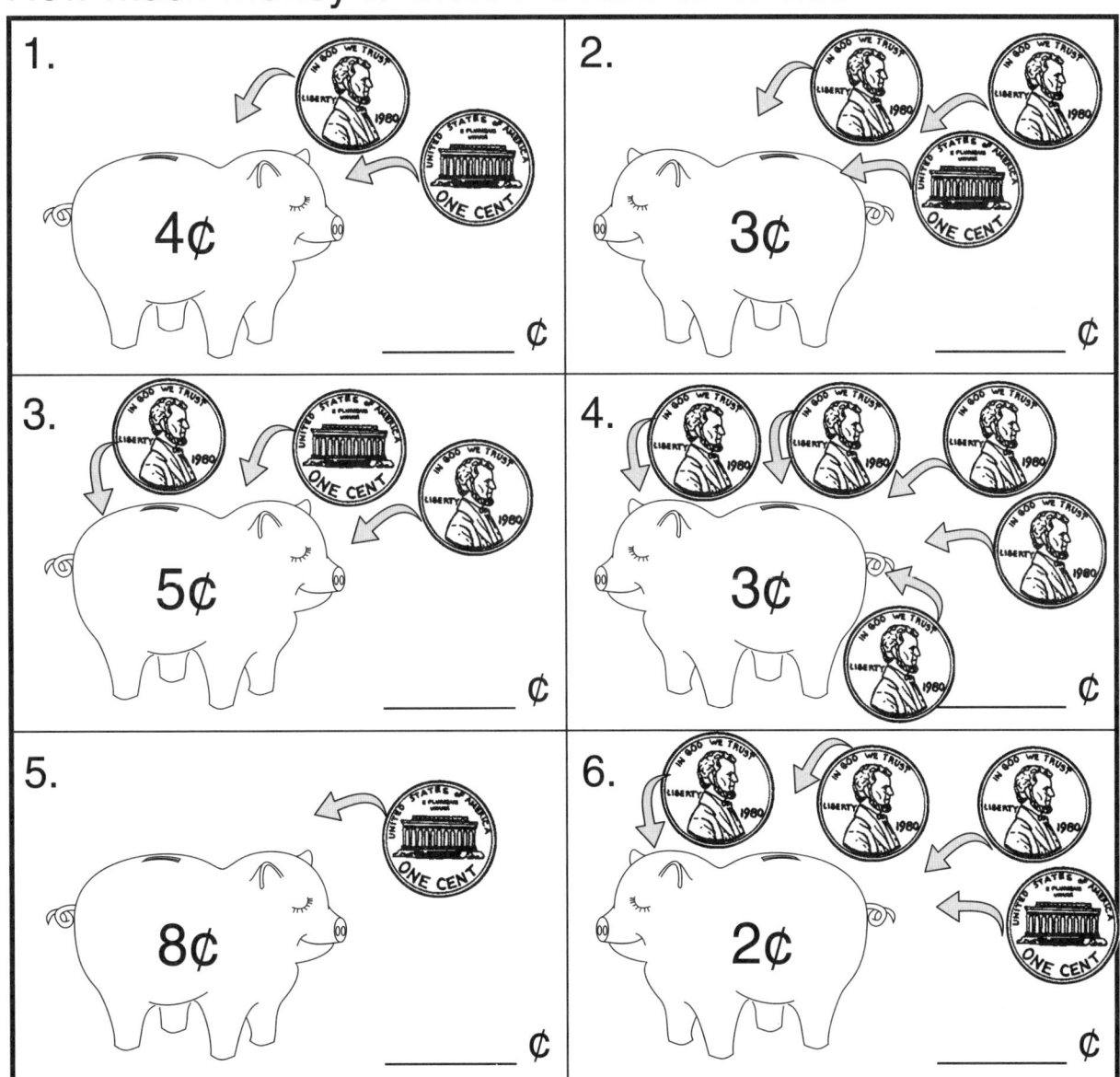

1. 4¢ _____ ¢

2. 3¢ _____ ¢

3. 5¢ _____ ¢

4. 3¢ _____ ¢

5. 8¢ _____ ¢

6. 2¢ _____ ¢

Name _____

Problem Solving: Guess and Check

Tell a story. Which fact goes with the picture?

1. Gum 1¢   Balloon 4¢
   ○ 4¢ + 1¢ = 5¢
   ○ 5¢ + 1¢ = 6¢

2. 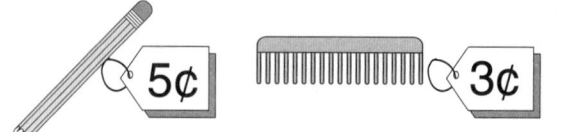 Pencil 5¢   Comb 3¢
   ○ 3¢ + 2¢ = 5¢
   ○ 5¢ + 3¢ = 8¢

3. 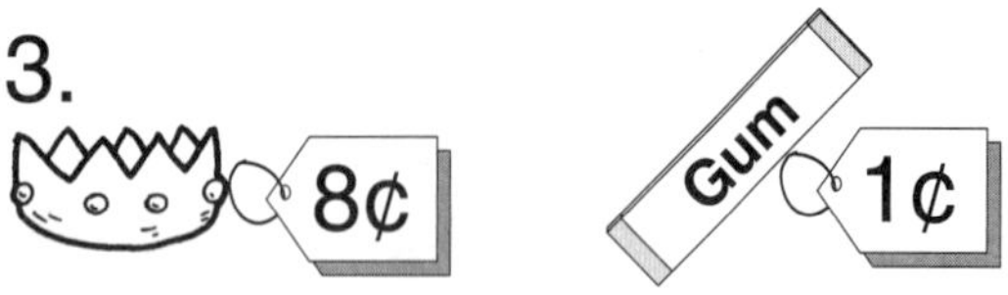 Crown 8¢   Gum 1¢
   ○ 8¢ + 1¢ = 9¢
   ○ 9¢ + 1¢ = 10¢

4.  Apple 7¢   Lollipop 2¢
   ○ 7¢ + 2¢ = 9¢
   ○ 9¢ + 1¢ = 10¢

5. 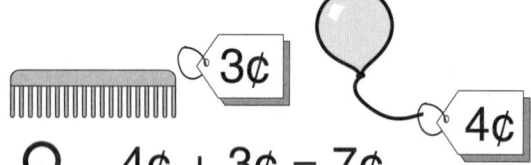 Comb 3¢   Balloon 4¢
   ○ 4¢ + 3¢ = 7¢
   ○ 7¢ + 3¢ = 10¢

6.  Lollipop 2¢   Whistle 6¢
   ○ 6¢ + 2¢ = 8¢
   ○ 8¢ + 2¢ = 10¢

7. You spent 8¢. What did you buy? Ring.

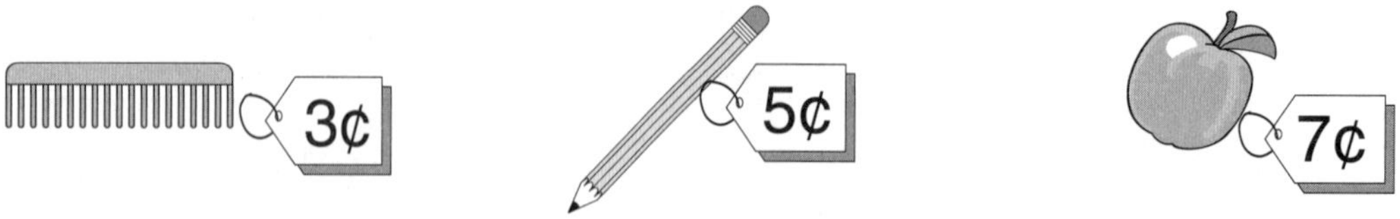

   Comb 3¢   Pencil 5¢   Apple 7¢

8. You spent 9¢. What did you buy? Ring.

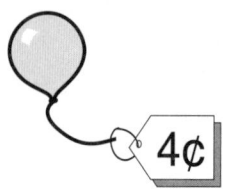

   Balloon 4¢   Pencil 5¢   Whistle 6¢

Name _____

Chapter 3 Review

1. How many in all? Write a number sentence.

   _____ _____ _____

   _____ _____ _____

2. Write the sentence. Add.

   ☐ _____ ☐ _____

3. Write the sentence. Add.

   ☐ _____ ☐ _____

4. Find the sums.

     2   4   5   3   1   3
   +2   +0   +1   +1   +4   +3

5. Find the sums.

     6   3   3   7   9   4
   +2   +3   +4   +3   +1   +5

6. Find the sums.

   0 + 5 = _____   8 + 2 = _____   3 + 4 = _____

Name _____

Chapter 3 Review

7. Finish the table.

Add 3	
6	
3	
7	

8. Ring a fact for the number.

9

7 + 3

3 + 4

4 + 5

9. How much?

_____ ¢

10. How much money? Count on to add.

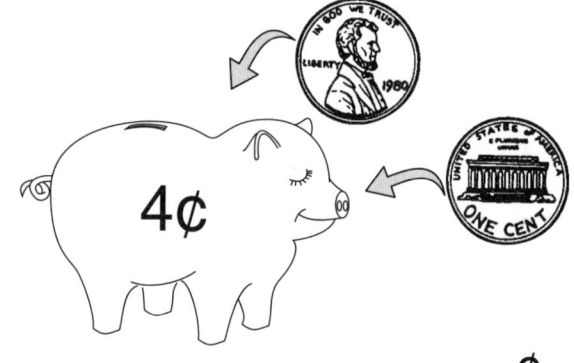

4¢

_____ ¢

11. Which fact goes with the picture?

Gum 1¢   8¢

○ 8¢ + 1¢ = 9¢
○ 9¢ + 1¢ = 10¢

12. Which fact goes with the picture?

6¢   2¢

○ 6¢ + 2¢ = 8¢
○ 8¢ + 2¢ = 10¢

4
Differences to 10

Name _____

Fill in the blanks

1.

____ in all    ____ going away    ____ left

2.

____ in all    ____ going away    ____ left

3.

____ in all    ____ going away    ____ left

4.

____ in all    ____ going away    ____ left

Name _____

One Less
Tell the story.

1.

2.

3.

Name _____

Two Less and Three Less
Tell the story. Write the numbers.

1.

2.

3.

4.

94
© Math Teachers Press, Inc., Reproduction by any means is strictly prohibited.

Name _____

Minus Sign and Equal Sign
Tell the story.

1. 4 ☐ 2 ☐ 2

2. ___ ☐ ___ ☐ ___

3. ___ ☐ ___ ☐ ___

4. ___ ☐ ___ ☐ ___

Name _____

Problem Solving: Writing a Number Sentence

1. _____ ☐ _____ ☐ _____

2. _____ ☐ _____ ☐ _____

3. _____ ☐ _____ ☐ _____

4. _____ ☐ _____ ☐ _____

5. Tell a story or draw a picture about 5 – 2.

_____

_____

Name _____

Problem Solving: Use a Model, Draw a Picture

_____ ☐ _____ ☐ _____

Build the larger number.   Take away.   Number left.

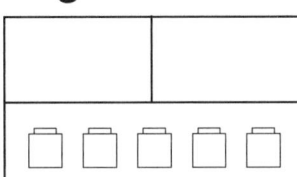

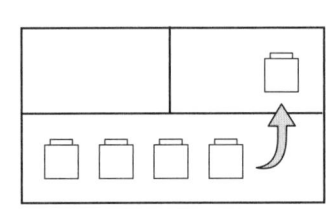

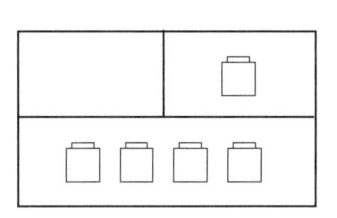

_____ − _____ = _____

Draw a picture. Subtract.

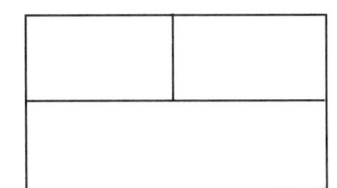

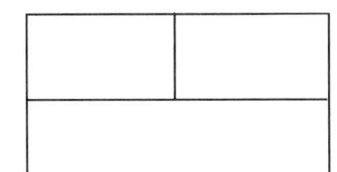

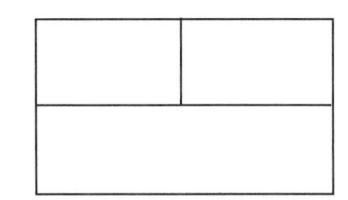

1.

                                                        4

2.

                                                        2

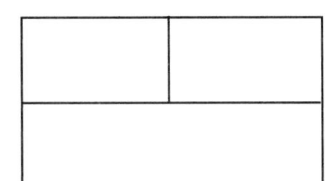

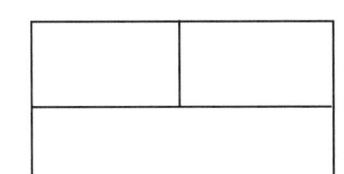

Name _____

Tell the Story

Number sentence: _____

Tell a story. Subtract with the mat.

1.   5 − 1 = _____   6 − 5 = _____   5 − 3 = _____

2.   5 − 5 = _____   6 − 3 = _____   4 − 2 = _____

3.   6 − 4 = _____   5 − 0 = _____   3 − 3 = _____

4.   4 − 3 = _____   6 − 2 = _____   5 − 2 = _____

5.   5 − 4 = _____   4 − 4 = _____   3 − 0 = _____

© Math Teachers Press, Inc., Reproduction by any means is strictly prohibited.

Name _____

Recording Another Way: Vertical

4 children. 1 leaves.

4 – 1 = _____

4
– 1
___

Tell the story. Write the sentence in 2 ways.

1.

3 – 2 = _____

3
– 2
___

2.

___ – ___ = ___

–
___

3.

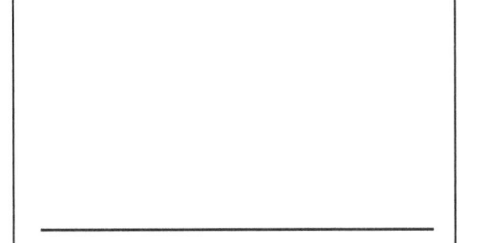

___ – ___ = ___

–
___

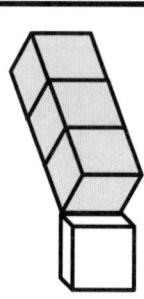

Name _____

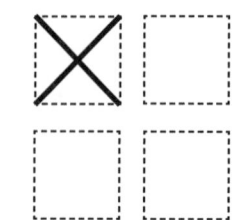

3 − 1 = ___

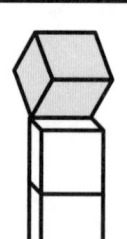

3
-1
___

Tell a story.
Write the number sentence in vertical format.

1.
3
-2
___

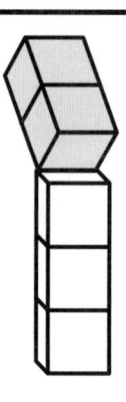

2.
− ___

3.
___

4.
___

5.
___

6.
___

100

Problem Solving

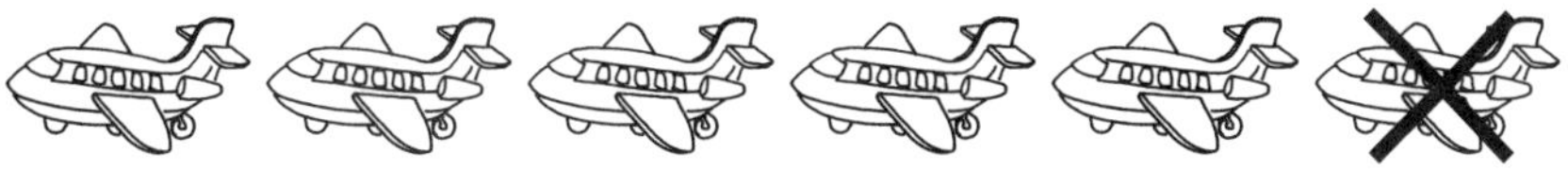

6 − =

Tell a story. Write the number sentence.

1. 5 − 2 = ___

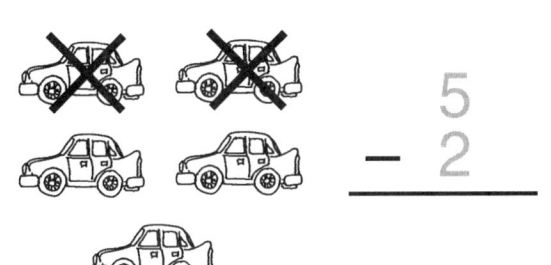

2. − = ___

3. − = ___

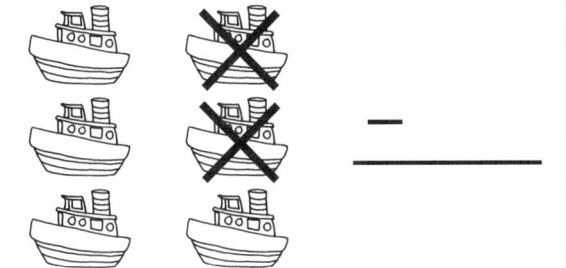

4. − = ___

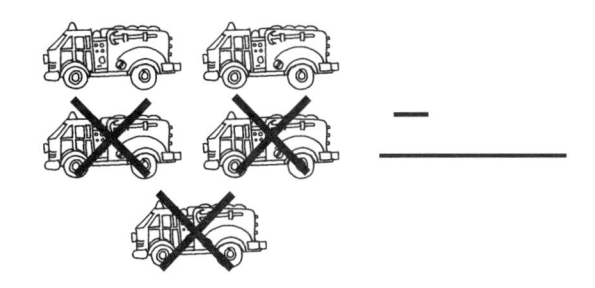

5. − = ___

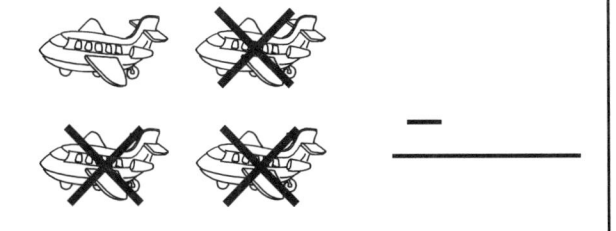

6. − = ___

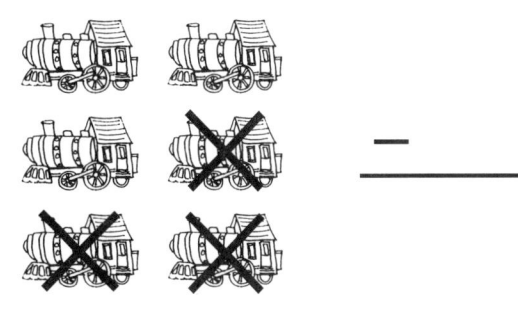

101

Name _____

Problem Solving: Mental Math with Zero

Start with 5. Eat 0. (none)

$$5 - 0 = 5$$

Start with 5 cookies. Eat them all.

$$5 - 5 = 0$$

Use the ▭▭ mat. Subtract.

1.   4   3   6   2   1   1
    −4   −0   −6   −0   −1   −0

2.   3   4   5   6   5   6
    −3   −0   −5   −0   −0   −6

3. 6 − 0 = _____   2 − 2 = _____   4 − 2 = _____

4. 5 − 5 = _____   3 − 3 = _____   6 − 6 = _____

Name _____

Problem Solving: Mental Math by Counting Back

$$\begin{array}{r} 6 \\ -2 \\ \hline \end{array}$$

Start at 6. Count back 2 steps. What is the landing point?

Subtract by counting back.

1.
$$\begin{array}{r} 4 \\ -1 \\ \hline \end{array} \quad \begin{array}{r} 5 \\ -2 \\ \hline \end{array} \quad \begin{array}{r} 6 \\ -3 \\ \hline \end{array} \quad \begin{array}{r} 3 \\ -3 \\ \hline \end{array} \quad \begin{array}{r} 5 \\ -3 \\ \hline \end{array} \quad \begin{array}{r} 3 \\ -1 \\ \hline \end{array}$$

2.
$$\begin{array}{r} 5 \\ -3 \\ \hline \end{array} \quad \begin{array}{r} 6 \\ -2 \\ \hline \end{array} \quad \begin{array}{r} 5 \\ -1 \\ \hline \end{array} \quad \begin{array}{r} 2 \\ -2 \\ \hline \end{array} \quad \begin{array}{r} 4 \\ -3 \\ \hline \end{array} \quad \begin{array}{r} 3 \\ -2 \\ \hline \end{array}$$

3.
$$\begin{array}{r} 6 \\ -1 \\ \hline \end{array} \quad \begin{array}{r} 4 \\ -2 \\ \hline \end{array} \quad \begin{array}{r} 6 \\ -3 \\ \hline \end{array} \quad \begin{array}{r} 5 \\ -0 \\ \hline \end{array} \quad \begin{array}{r} 2 \\ -1 \\ \hline \end{array} \quad \begin{array}{r} 4 \\ -0 \\ \hline \end{array}$$

4. 5 − 1 = _____   6 − 3 = _____   4 − 2 = _____

5. 4 − 1 = _____   5 − 3 = _____   6 − 2 = _____

Name _____

Problem Solving: Mental Math by Counting Up

Here are two different ways to subtract the same problem. Which is easier?

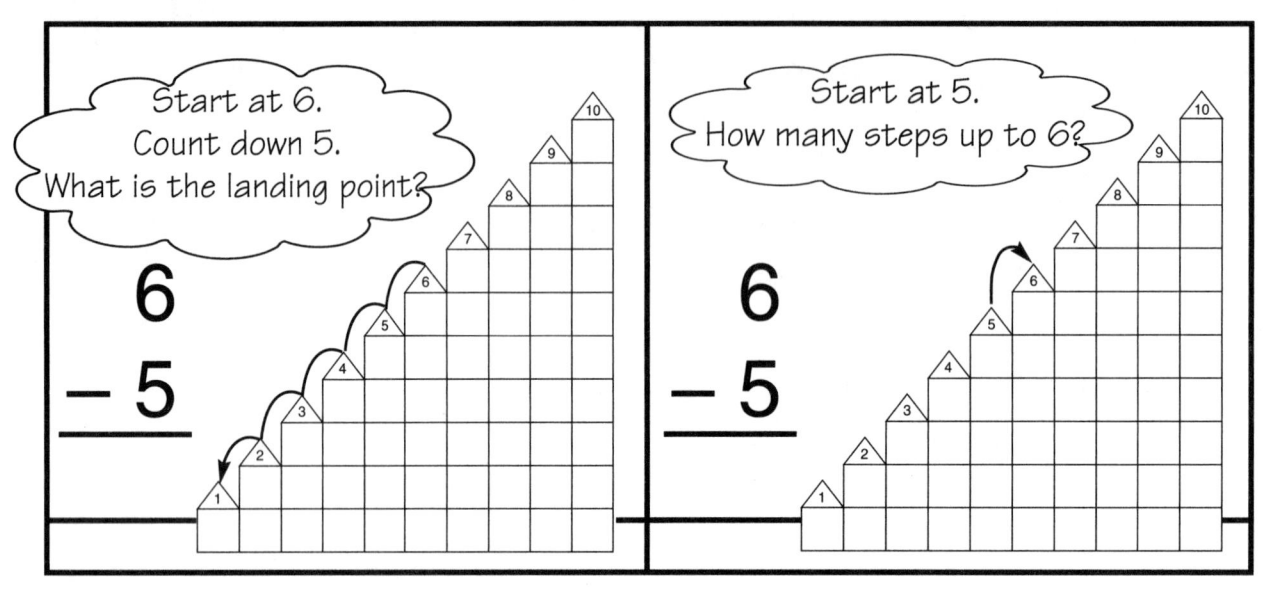

Start at 6. Count down 5. What is the landing point?

6
− 5
___

Start at 5. How many steps up to 6?

6
− 5
___

Subtract. Think of counting up by 1, 2 or 3.

1.   5   6   5   3   6   6
    − 4   − 3   − 3   − 2   − 5   − 4

2.   6   6   4   5   3   5
    − 3   − 5   − 3   − 3   − 2   − 4

Subtract.

3.   5   6   6   4   6   5
    − 5   − 2   − 3   − 0   − 6   − 2

4. 6 − 4 = _____   5 − 3 = _____   3 − 0 = _____

5. 4 − 4 = _____   6 − 5 = _____   4 − 3 = _____

Name _____

Solve. ✏️.

0 = yellow   3 = green
1 = red   4 = orange
2 = blue   5 = brown
              6 = black

3−3=
3−2
6−0
6−6
4−2=
1−1
6−1=
6−5
5−1
5−4=
5−5
5−0=
6−4
3−0=
4−2 (shown as 3 −2)
4−1
5−2=
6−1
5−3
1−0=
4−3=
4−4
4−0
2−1=
6−2=

105
© Math Teachers Press, Inc., Reproduction by any means is strictly prohibited.

Name _____

No. Correct (28) _____

Review: Subtracting from Numbers up to 6

1.
6	5	4	3	2	3
−1	−3	−2	−0	−2	−2

2.
6	5	1	5	4	6
−0	−1	−1	−0	−1	−2

3.
5	4	3	0	4	6
−5	−0	−1	−0	−4	−3

4.
6	5	4	3	5	2
−4	−2	−3	−0	−4	−1

5.
6	4	1	6
−5	−3	−0	−6

Write the missing signs (+ or −).

2 ☐ 3 = 5   6 ☐ 2 = 4   5 ☐ 2 = 3

Name _____

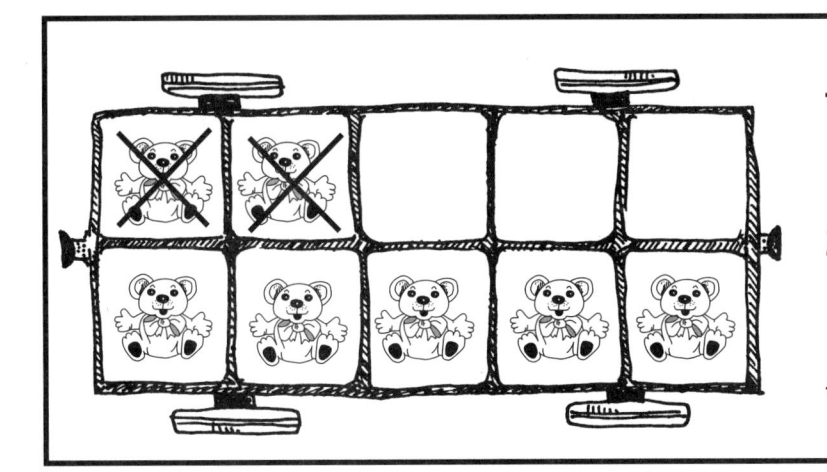

There are 7 🧸.

2 🧸 get off.

_____ are left.

1. There are 8 🧸.

   1 goes away.

   _____ are left.

2. There are 7 🧸.

   3 go away.

   _____ are left.

3. There are 9 🧸.

   1 leaves.

   _____ are left.

4. There are 8 🧸.

   2 get off.

   _____ are left.

5. There are 10 🧸.

   1 gets off.

   _____ are left.

6. There are 9 🧸.

   2 get off.

   _____ are left.

107
© Math Teachers Press, Inc., Reproduction by any means is strictly prohibited.

Name _____

Subtracting on a Number Line

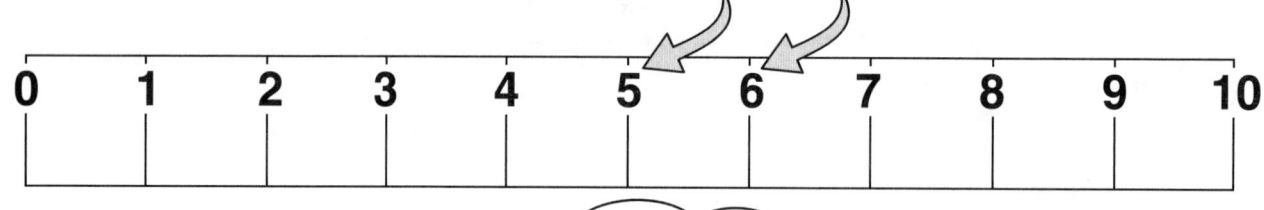

0 1 2 3 4 5 6 7 8 9 10

7 − 2 = ___   *Start at 7. Jump back 2 steps. 7, 6, 5*   7 − 2 = 5

1.
8, 7, 6　　*9, 8, 7, 6*　　*7, 6*　　*10, 9, 8*
8　　　9　　　7　　　10
−2　　−3　　−1　　−2

2.　7　　　8　　　10　　　9
　−3　　−1　　−1　　−2

3.　8　　　7　　　9　　　10
　−3　　−2　　−1　　−3

4.　6　　　5　　　6　　　4
　−3　　−0　　−2　　−4

Name _____

There are 7 🐻.

5 🐻 get off.

_____ left.

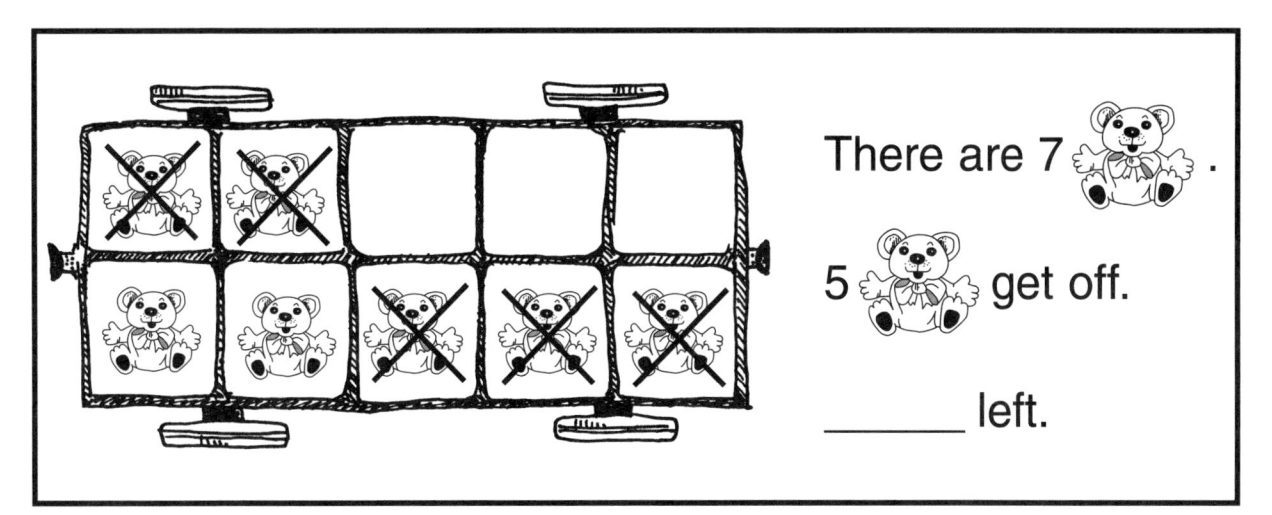

Subtract by drawing a picture.

1. There are 8 🐻.

   7 🐻 get off.

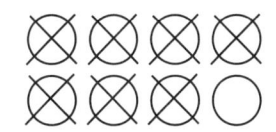

   ⊗⊗⊗⊗
   ⊗⊗⊗○

   _____ left.

2. There are 9 🐻.

   6 🐻 get off.

   ○○○
   ○○○
   ○○○

   _____ left.

3. There are 10 🐻.

   9 🐻 get off the train.

   _____ left.

4. There are 7 🐻.

   4 🐻 go away.

   _____ left.

5. There are 9 🐻.

   7 🐻 get off the train.

   _____ left.

6. There are 8 🐻.

   5 🐻 get off the train.

   _____ left.

Name _____

Subtracting on a Number Line

0 1 2 3 4 5 6 7 8 9 10

7 − 5 = ___   *Start at 5. Count up to 7. 5, 6, 7... 2 steps up to 7*   7 − 5 = 2

Subtract.

5, 6, 7, 8 — 3 steps up to 8   *6, 7 — 1 step up*   *8, 9, 10 — 2 steps up*   *6, 7, 8 — 2 steps up*

1. 8   7   10   8
   −5   −6   −8   −6

2. 9   10   7   9
   −8   −7   −5   −6

3. 7   8   9   8
   −4   −7   −7   −5

4. 8   10   9   8
   −6   −9   −6   −7

110
© Math Teachers Press, Inc., Reproduction by any means is strictly prohibited.

Name _____

There are 8 🐑 to pet.

4 🐑 go away.

_____ left.

Subtract by drawing a picture.

1. There are 9 🐑.

   4 🐑 go away.

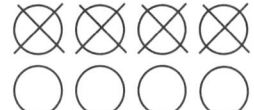

   ⊗⊗⊗⊗
   ○○○○
   ○   _____ left.

2. There are 10 🐑.

   4 🐑 go away.

   ○○○○
   ○○○
   ○○○   _____ left.

3. There are 10 🐴.

   5 🐴 hide.

   _____ left.

4. There are 10 🐴.

   4 🐴 hide.

   _____ left.

5. There are 8 🐤.

   8 🐤 go away.

   _____ left.

6. There are 10 🐤.

   6 🐤 go away.

   _____ left.

Name _____

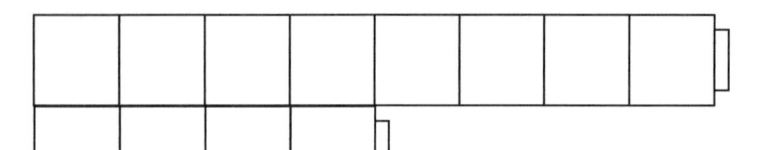

8 − 4 =

How many cubes do you add to the "4" train to make the "8" train?

8 − 4 = _____

Use cubes to find the difference.

1. 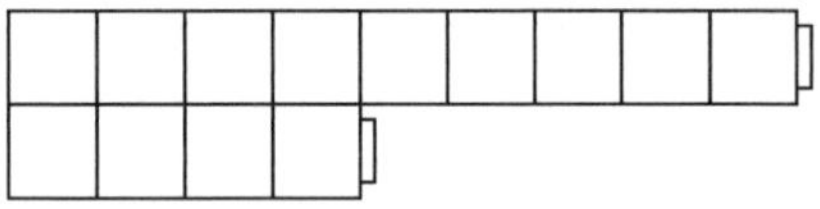 9 − 4 = _____

2. 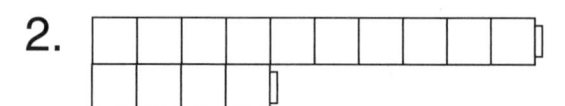 10 − 4 = _____

3. 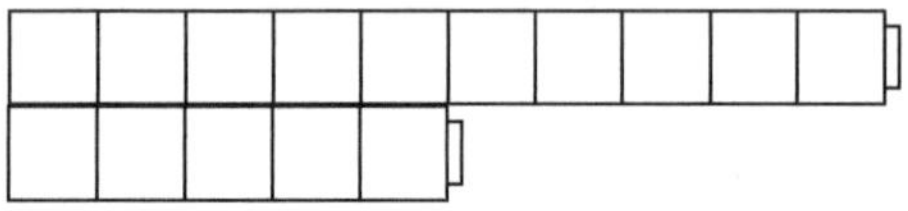 10 − 5 = _____

4. 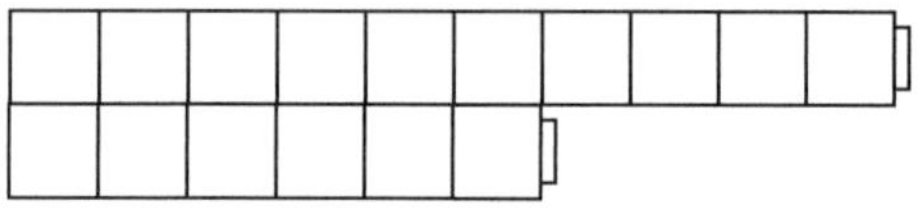 10 − 6 = _____

5. 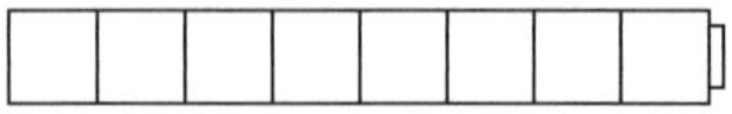 8 − 0 = _____

6. 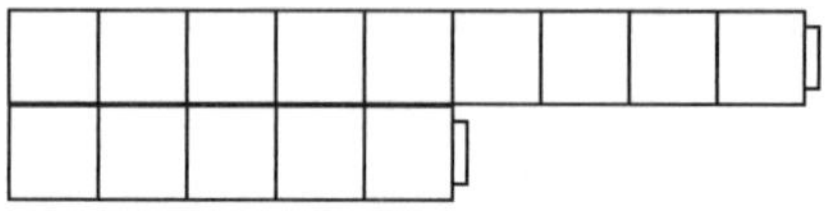 9 − 5 = _____

7.
10	9	7	10	8	10
− 6	− 5	− 0	− 5	− 8	− 4

8.
9	8	8	7	9	9
− 9	− 4	− 0	− 7	− 4	− 0

Name _____

Relating Addition and Subtraction

$4 + 3 = 7$ $7 - 3 = 4$

Write an addition fact and a subtraction fact for each picture.

1.
$$\begin{array}{r} 3 \\ + 2 \\ \hline 5 \end{array} \qquad \begin{array}{r} 5 \\ - 2 \\ \hline 3 \end{array}$$

2.
+ _____   − _____

3.
+ _____   − _____

4.
+ _____   − _____

5.
+ _____   − _____

6.
+ _____   − _____

113

Name _____

Addition and subtraction are opposites.
One undoes the other.

Addition puts things together.	Subtraction takes things apart.
4 + 3 7 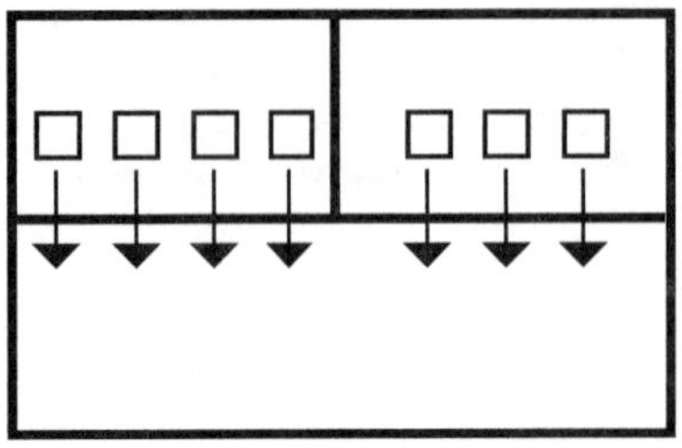	7 − 3 4 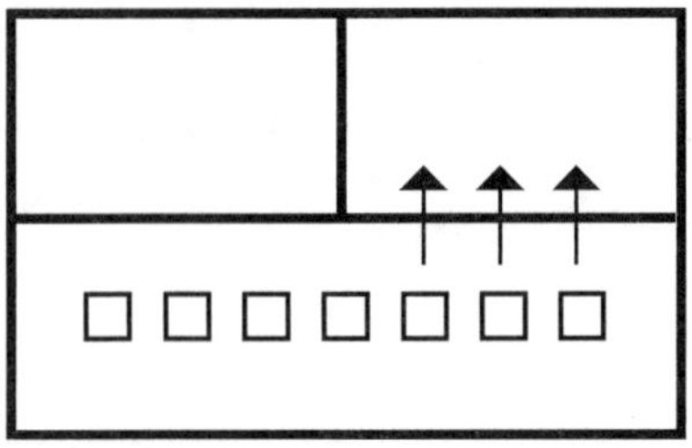

Write an addition fact and a subtraction fact for each picture.

1.

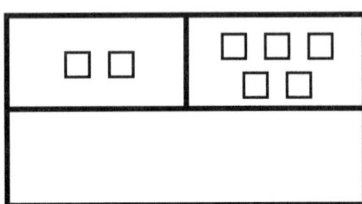

2
+ 5

7
− 5

2.

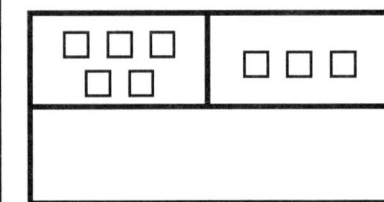

____ + ____   ____ − ____

3.

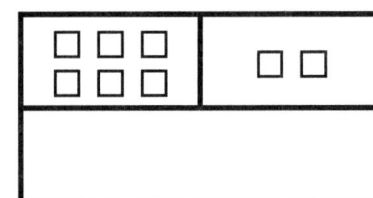

____ + ____   ____ − ____

4.

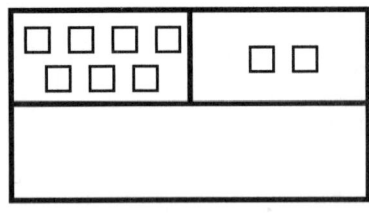

____ + ____   ____ − ____

5.

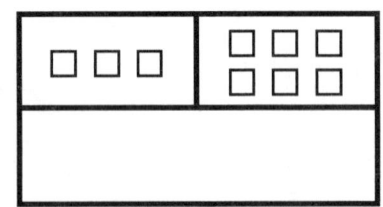

____ + ____   ____ − ____

6.

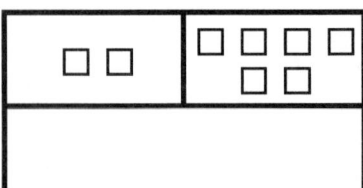

____ + ____   ____ − ____

Using a Calculator: Counting Back to Subtract

1. Use your calculator to count back 1.

   [on/c] [on/c] [4] [-] [1] [=] _____ [=] _____

2. Count back 1.

   [on/c] [on/c] [5] [-] [1] [=] _____ [=] _____

3. Count back 2.

   [on/c] [on/c] [3] [-] [2] [=] _____

4. Count back 3.

   [on/c] [on/c] [5] [-] [3] [=] _____

5. Count back 2.

   [on/c] [on/c] [6] [-] [2] [=] [=] [=] _____

Name _____

Differences to 10   No. Correct (36) _____

1.   7   9   9   7   7   10
    −2   −9   −1   −0   −1   −9

2.   10   9   8   9   10   9
    −8   −4   −8   −5   −7   −3

3.   10   9   8   8   9   7
    −5   −7   −2   −0   −6   −7

4.   9   8   10   8   9   8
    −8   −4   −3   −6   −0   −5

5.   9   10   8   10   7   7
    −2   −1   −7   −2   −5   −4

6.   10   8   8   7   7   10
    −6   −3   −1   −6   −3   −4

© Math Teachers Press, Inc., Reproduction by any means is strictly prohibited.

Name _____

Subtracting Pennies

Pedro had 6¢.
He spent 4¢.
How much is left?

6¢
− 4¢
2¢

Tell a story.
How much money is left?

1. 5¢
   − 2¢
   ___ ¢

2. 6¢
   − 3¢
   ___ ¢

3. ___ ¢
   − ___ ¢
   ___ ¢

4. ___ ¢
   − ___ ¢
   ___ ¢

Draw a picture. Solve.

5. Linda had 7¢.

   She spent 5¢.

   _____ ¢ left.

6. Carl had 10¢.

   He spent 5¢.

   _____ ¢ left.

7. George had 8¢.

   He lost 4¢.

   _____ ¢ left.

8. Gretta had 7¢.

   She gave away 3¢.

   _____ ¢ left.

Name _____

Kim had 7¢.
She spent 2¢.

7¢

Count back.
7, **6, 5**

How much is left? _____ ¢

How much money is left?
Count back.

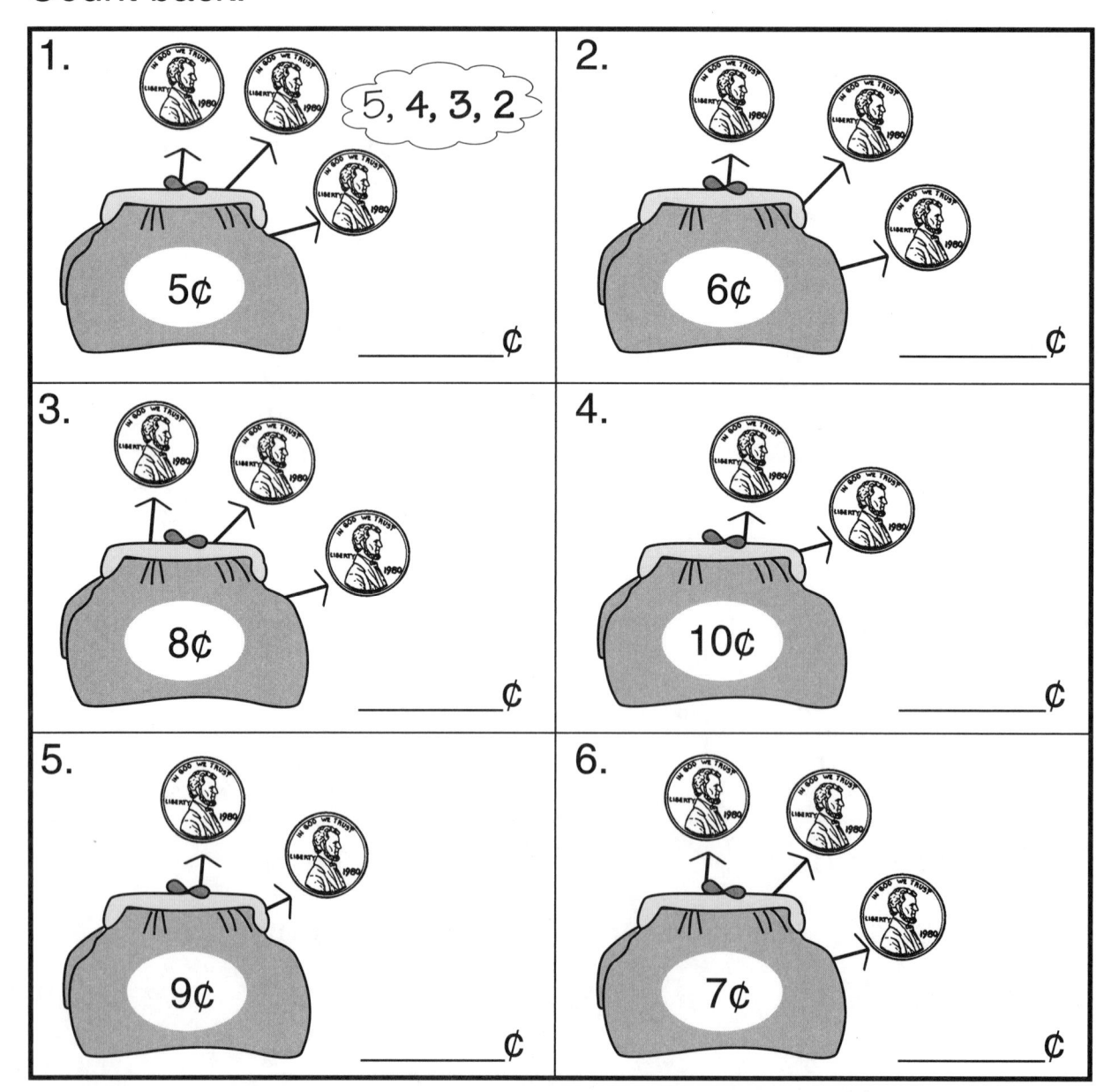

1. 5¢   5, 4, 3, 2   _____ ¢

2. 6¢   _____ ¢

3. 8¢   _____ ¢

4. 10¢ _____ ¢

5. 9¢   _____ ¢

6. 7¢   _____ ¢

Name _____

Problem Solving: Choose the Correct Number Sentence

Choose the correct number sentence.

○ 4 + 3 = 7
○ 7 − 3 = 4

○ 4 + 2 = 6
○ 6 − 2 = 4

Tell the story. Choose the correct number sentence.

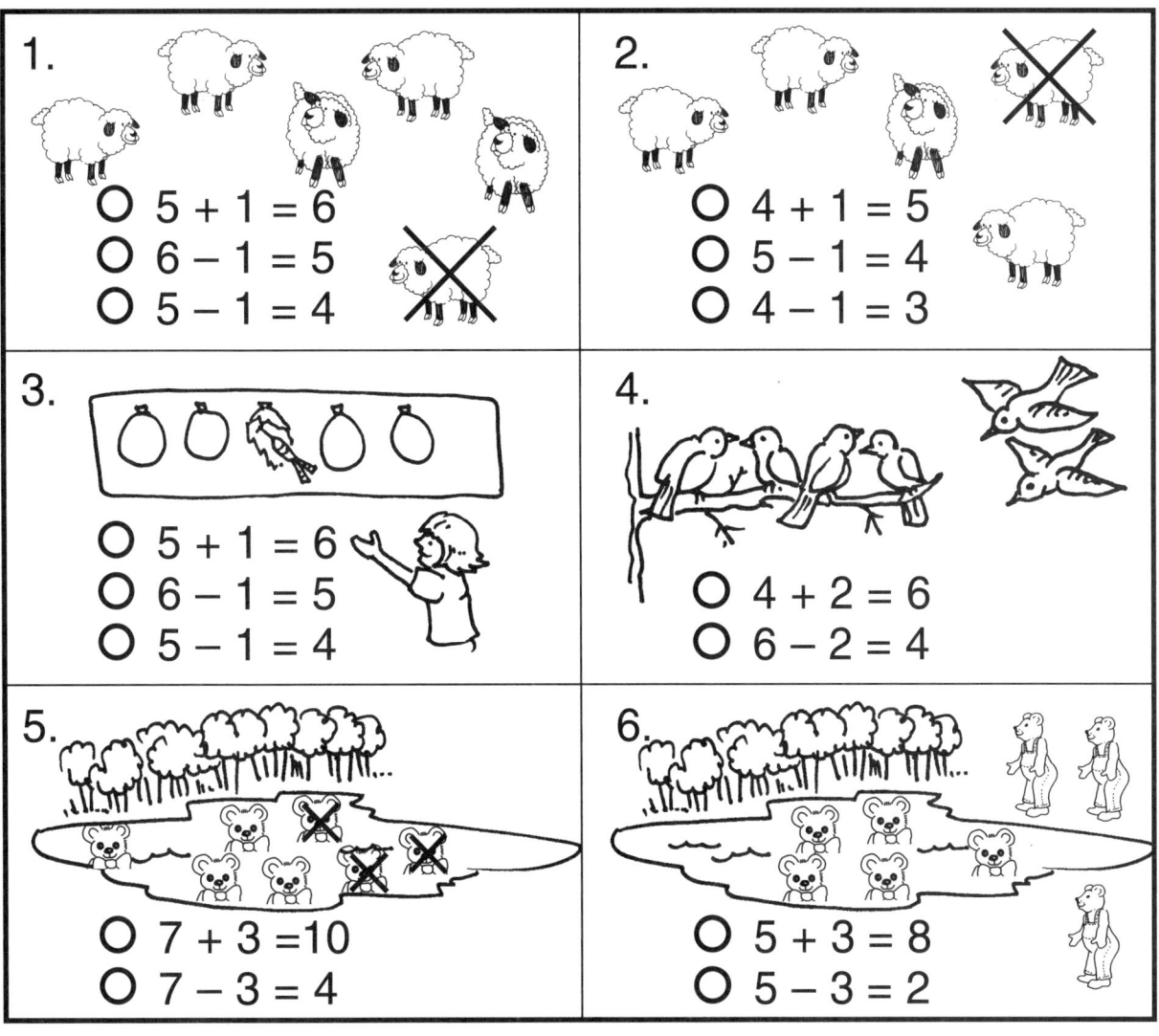

1.
○ 5 + 1 = 6
○ 6 − 1 = 5
○ 5 − 1 = 4

2.
○ 4 + 1 = 5
○ 5 − 1 = 4
○ 4 − 1 = 3

3.
○ 5 + 1 = 6
○ 6 − 1 = 5
○ 5 − 1 = 4

4.
○ 4 + 2 = 6
○ 6 − 2 = 4

5.
○ 7 + 3 = 10
○ 7 − 3 = 4

6.
○ 5 + 3 = 8
○ 5 − 3 = 2

119

Name _____

Favorite Farm Animals

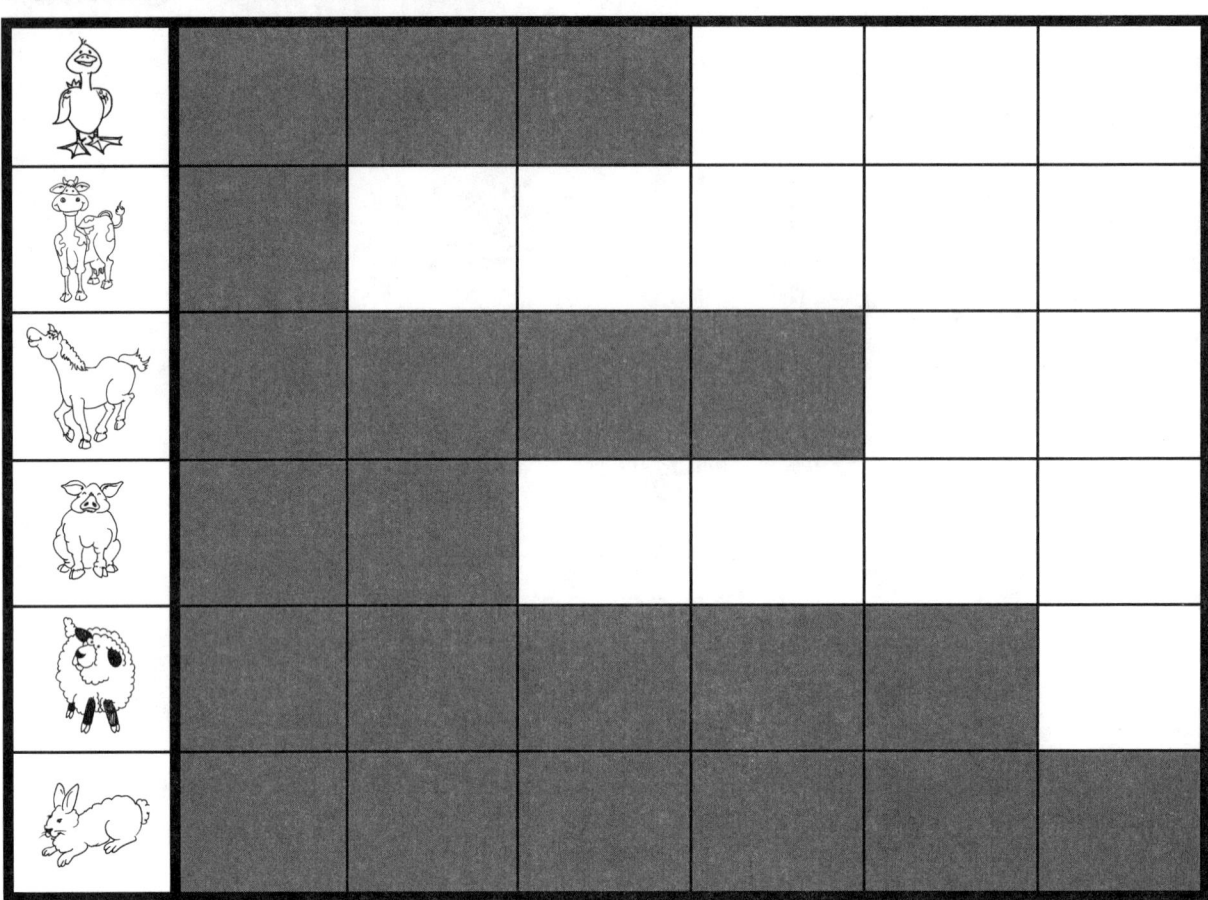

1. How many 🦆 ? _____ 
2. How many 🐷 ? _____ 
3. How many 🦆 and 🐷 ? _____ 
4. How many 🐴 ? _____ 
5. Ring the favorite animal.   
6. Ring the least favorite animal.

120
© Math Teachers Press, Inc., Reproduction by any means is strictly prohibited.

Name _____

Chapter 4 Review

1. How many are left? Write a number sentence.

   ☐ _____ ☐ _____

2. Write the sentence. Subtract.

   _____ − _____ = _____

3. Write the sentence. Subtract.

   −
   _____

4. Find the differences.

   $\begin{array}{r}5\\-1\\\hline\end{array}$ $\begin{array}{r}6\\-2\\\hline\end{array}$ $\begin{array}{r}4\\-0\\\hline\end{array}$ $\begin{array}{r}6\\-3\\\hline\end{array}$ $\begin{array}{r}5\\-2\\\hline\end{array}$ $\begin{array}{r}3\\-3\\\hline\end{array}$

5. Subtract.

   $\begin{array}{r}7\\-2\\\hline\end{array}$ $\begin{array}{r}9\\-1\\\hline\end{array}$ $\begin{array}{r}7\\-0\\\hline\end{array}$ $\begin{array}{r}10\\-2\\\hline\end{array}$ $\begin{array}{r}8\\-8\\\hline\end{array}$ $\begin{array}{r}9\\-3\\\hline\end{array}$

6. Find the differences.

   4 − 2 = _____   5 − 3 = _____   10 − 5 = _____

Name _____

7. Subtract by drawing a picture.

There are 7 🐻.

4 🐻 go away.

_____ left.

8. Ring a fact for the number in the box.

| 3 |

6 − 2

5 − 3

5 − 2

9. How much money is left?

9¢
− 5¢
___ ¢

10. How much money is left? Count back to subtract.

6¢

_____ ¢

11. Which fact goes with the picture?

○ 4 + 1 = 5
○ 5 − 1 = 4
○ 4 − 1 = 3

12. Which fact goes with the picture?

○ 4 + 2 = 6
○ 6 − 2 = 4
○ 4 − 2 = 2

122
© Math Teachers Press, Inc., Reproduction by any means is strictly prohibited.

5
Numbers to 100

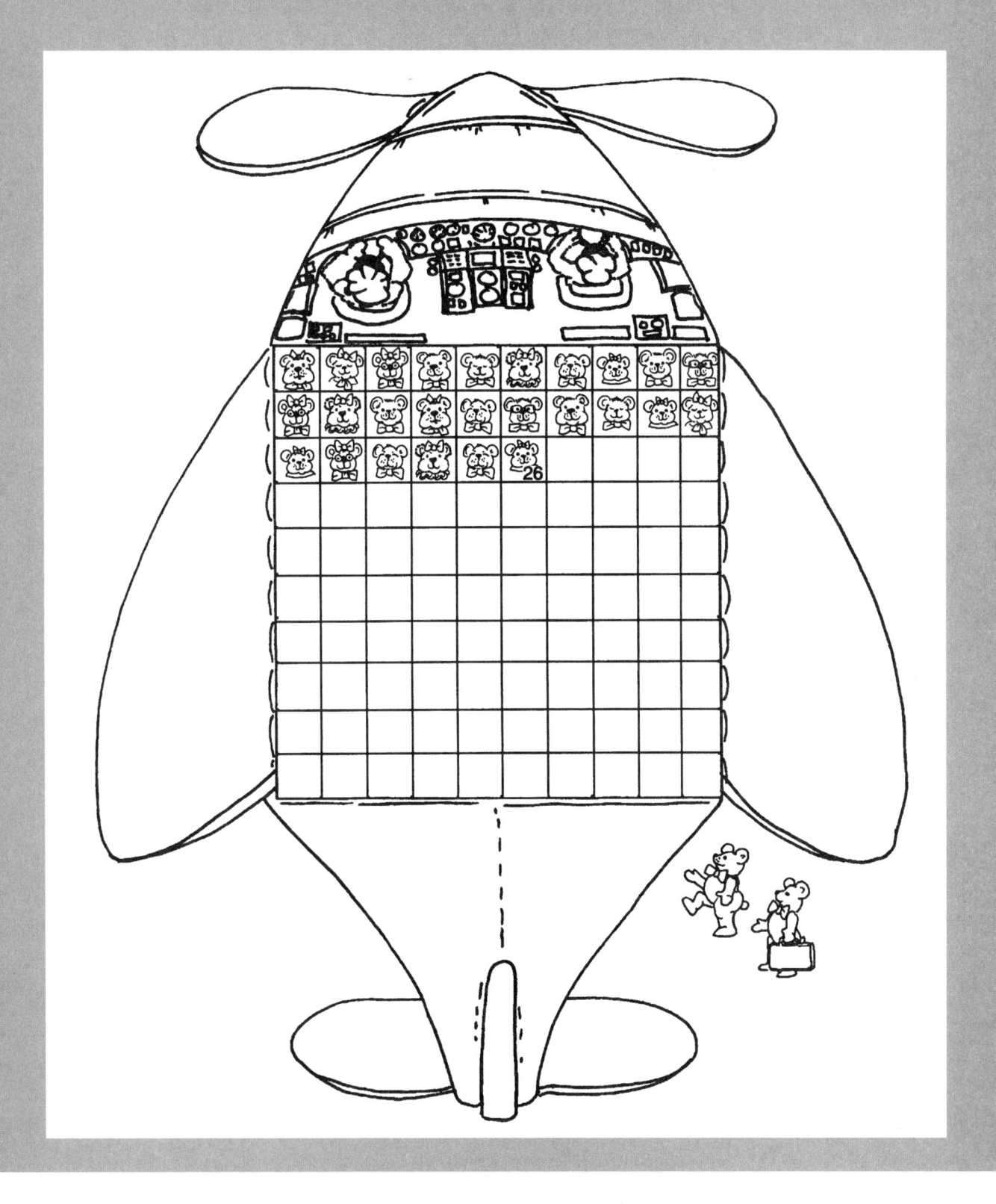

Name _____

Draw a line to match the number and word name.

1	(3 beans)		four
2	(4 beans)		one
3	(1 bean)		six
4	(2 beans)		two
5	(7 beans)		five
6	(5 beans)		three
7	(6 beans)		nine
8	(9 beans)		eight
9	(10 beans)		eleven
10	(10 beans) (2 beans)		ten
11	(8 beans)		twelve
12	(10 beans) (1 bean)		seven

124

Name _____

More Bears on the Train

14 bears got on the train. How many full cars of 10? How many left over?

1 full car of 10   4 left over in the next car

1 ten 4 ones = 14

1. 13 bears got on the train. Draw a picture of the bears.

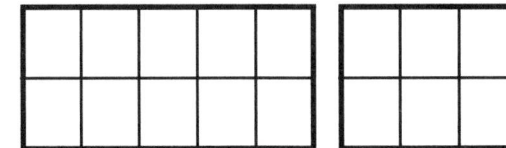

2. 16 bears got on the train. Draw a picture of the bears.

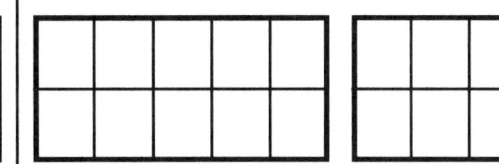

____ full cars ____ left over

13 = ____ tens ____ more

____ full cars ____ left over

16 = ____ tens ____ more

Draw a picture. Ring groups of 10.

3. 15 = ____ tens ____ more

(x x x x x)   x x x x x
(x x x x x)

4. 12 = ____ tens ____ more

5. 11 = ____ tens ____ more

6. 10 = ____ tens ____ more

Name _____

Count and ring groups of 10. Write how many.

1.

_____ tens _____ ones = 13

2.

_____ tens _____ ones = 14

3.

_____ tens _____ ones =

4.

_____ tens _____ ones =

5.

_____ tens _____ ones =

Tens and Ones

17 bears get on the plane. How many full rows of 10?
How many bears in the next row?

_____ full rows of 10 _____ ones in next row

17= _____ ten _____ ones

Draw a picture of the bears on the plane.

1. 18 bears get on the plane. How many full rows of 10? How many bears in the next row?

   _____ full rows of tens _____ ones bears

   18= _____ ten _____ ones

2. 20 bears get on the plane. How many full rows of 10? How many bears in the next row?

   _____ full rows of tens _____ ones bears

   20= _____ ten _____ ones

3. 19 bears get on the plane. How many full rows of 10? How many bears in the next row?

   _____ full rows of tens _____ ones bears

   19= _____ ten _____ ones

Name _____

Count. Ring groups of 10. Write how many.

1.

_____ tens _____ ones = 18

2.

_____ tens _____ ones = _____

3.

_____ tens _____ ones = _____

4.

_____ tens _____ ones = _____

5.

_____ tens _____ ones = _____

Name _____

> 10 ones is the same as... ...1 ten.

Ring groups of tens. Write the numeral in the blank.

1. _____ tens _____ ones

   _____

2. _____ tens _____ ones

   _____

3. _____ tens _____ ones

   _____

4. _____ tens _____ ones

   _____

5. _____ tens _____ ones

   _____

Name _____

Numbers to 20

[ten rod]	10	ten
[ten rod] [1 cube]	11	eleven
[ten rod] [2 cubes]	12	twelve
[ten rod] [3 cubes]	13	thirteen
[ten rod] [4 cubes]	14	fourteen
[ten rod] [5 cubes]	15	fifteen
[ten rod] [6 cubes]	16	sixteen
[ten rod] [7 cubes]	17	seventeen
[ten rod] [8 cubes]	18	eighteen
[ten rod] [9 cubes]	19	nineteen
[two ten rods]	20	twenty

Name _____

Comparing Numbers to 10

▭▭▭▭▭▭ is more than ▭▭▭▭	▭▭ is less than ▭▭▭▭
7 is greater than 5	2 is less than 4
7 > 5	2 < 4

Ring the correct words and sign.

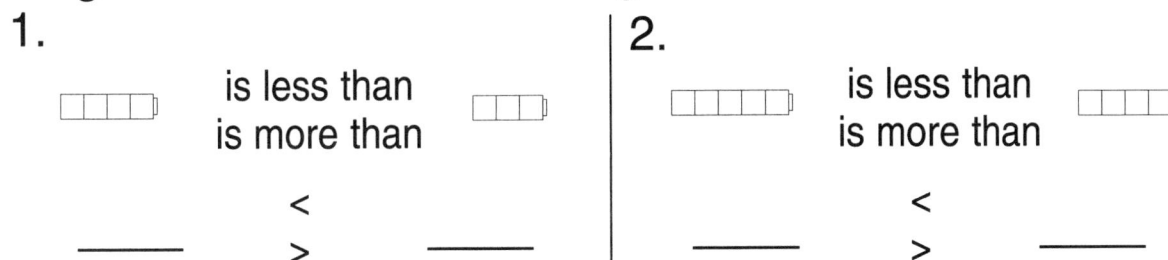

1. ▭▭▭▭ is less than ▭▭▭
           is more than
       ___ < ___
            >

2. ▭▭▭▭▭ is less than ▭▭▭▭
             is more than
         ___ < ___
              >

3. [7] is less than <   [4]
         is more than >

4. [6] is less than <   [8]
         is greater than >

<, >, = ? Write the correct sign in the circle.

5. 4 ◯ 6   5 ◯ 4   6 ◯ 6

6. 8 ◯ 7   9 ◯ 10   10 ◯ 8

7. 5 ◯ 5   6 ◯ 8   5 ◯ 6

8. 7 ◯ 9   10 ◯ 10   8 ◯ 6

Remember, open my mouth to the bigger number

Name ____________________

Here is the number line from 0 to 10.
Where do the numerals belong?

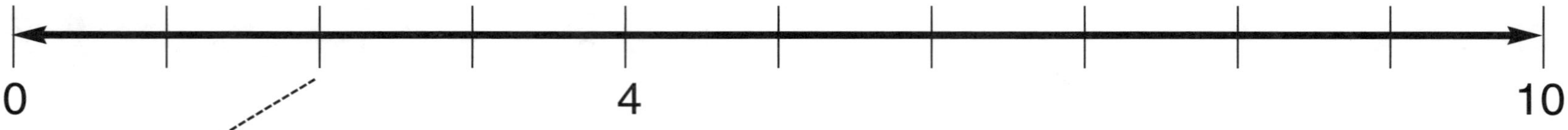

2 5 0 6 7 1 10 3 9 4 8

Here is the number line from 10 to 20.
Where do the numerals belong?

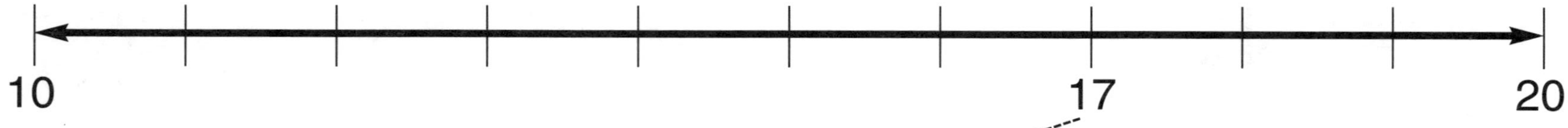

13 10 16 11 17 19 14 12 20 15 18

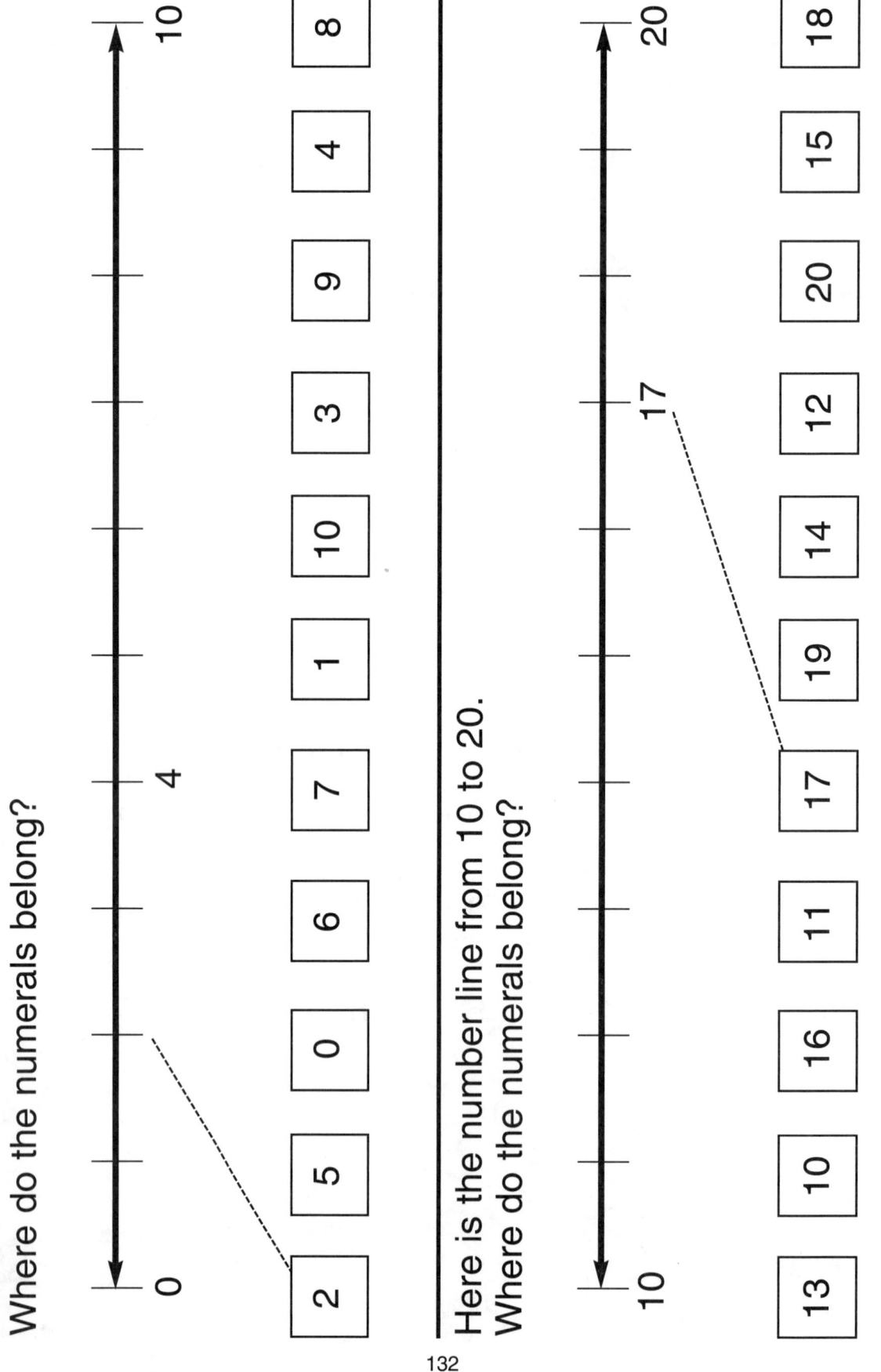

Name _____

1. Which number is **less**?

| 13 | 15 | 14 |
| 14 | 11 | 12 |

2. Which number is **greater**?

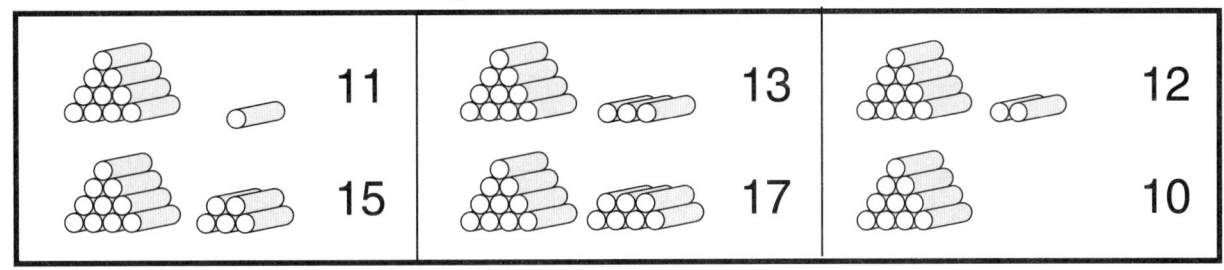

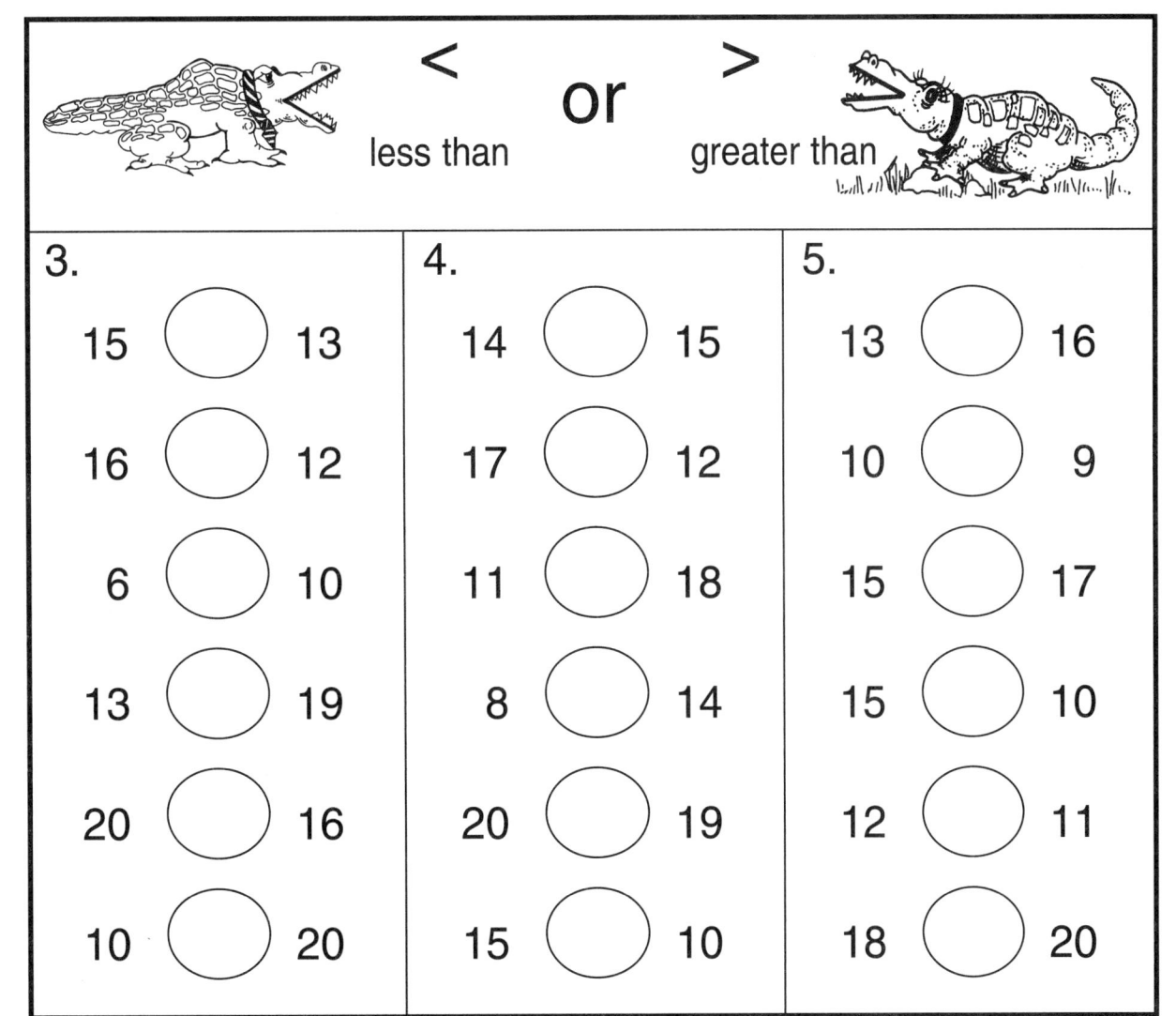

| 11 | 13 | 12 |
| 15 | 17 | 10 |

< or >
less than   greater than

3.
15 ◯ 13
16 ◯ 12
6 ◯ 10
13 ◯ 19
20 ◯ 16
10 ◯ 20

4.
14 ◯ 15
17 ◯ 12
11 ◯ 18
8 ◯ 14
20 ◯ 19
15 ◯ 10

5.
13 ◯ 16
10 ◯ 9
15 ◯ 17
15 ◯ 10
12 ◯ 11
18 ◯ 20

133
© Math Teachers Press, Inc. Reproduction by any means is strictly prohibited.

Name _____

Fill in each 🎟 .

13 14 15 16 17 18 19 20 21 22 23 24 25 26

10	11	12		
8				12
	16	17		
			16	17
	10	11		
			11	12
15	16			
9				13
17	18			
				11	12	

134
© Math Teachers Press, Inc. Reproduction by any means is strictly prohibited.

Name _____

Bears on a Plane

The bears are going on a plane. The bears board the plane and fill Row 1. Use your Plane Story Board and teddy bears.

1. What if 18 bears get on the plane?

   How many full rows? _____
   How many bears in the
   next row? _____

2. What if 30 bears get on an empty plane?

   _____   _____
   full rows   next row

3. What if 57 bears board an empty plane?

   _____ full rows   _____ next row

   _____ tens   _____ ones

4. What if 85 bears board an empty plane?

   _____ full rows   _____ next row

   _____ tens   _____ ones

5. Both groups have the same number of cubes.   Guess   Count
   Guess and count how many cubes are in group A. _____   _____
   Guess and count how many cubes are in group B. _____   _____
   Which way is easier to count the cubes? Why? _____

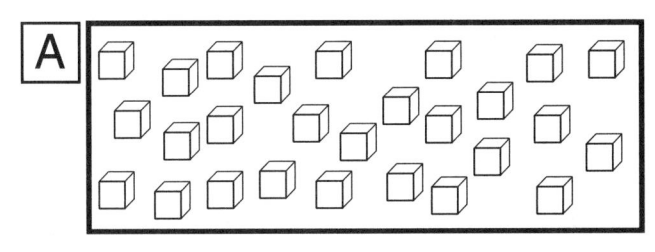

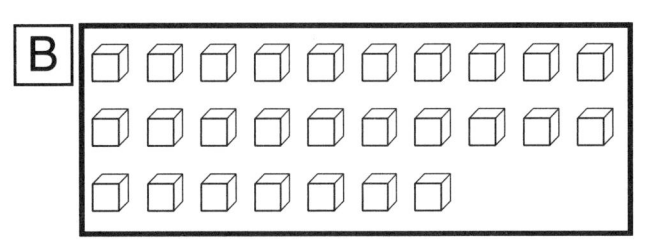

Name _____

Count to 100. Write the numbers on the tape.
The tape has been cut into 5 parts.

Tens	Ones
	1
	2
	3
	4
	5
	6
	7
	8
	9
1	0
1	1
1	5
2	0

Part 2

2	1
2	2
2	5
3	0
3	1
4	0

Part 3

4	1
4	8
4	9
5	5
5	9

Part 4

6	2
6	4
6	6
6	8
7	0
7	5

Part 5

8	5
9	0
9	9
1	0	0

Name _____

The Hundred Chart

Count by ones.
Write the missing numbers.

	2	3	4	5		7	8	9	
11		13	14		16	17	18		20
21	22		24	25		27		29	
31		33		35	36		38		40
41		43	44			47			50
51			54				58		60
61				65	66		68		
71			74			77			80
81				85					90
			94				98	99	

Name _____

1 2 3 4 5   7   9

Finish each row.

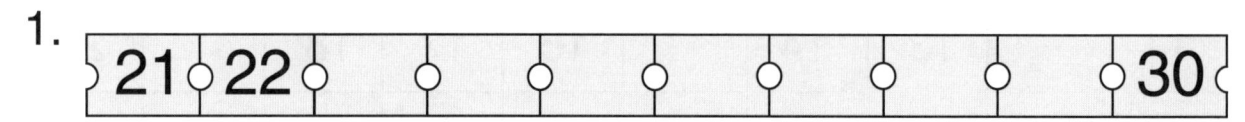

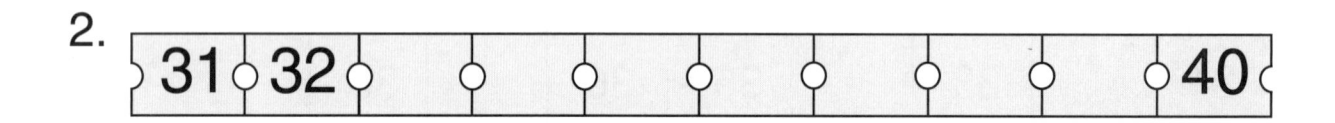

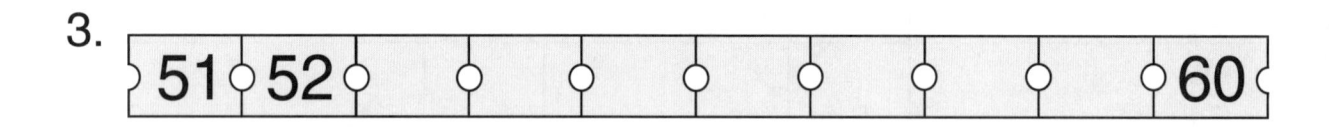

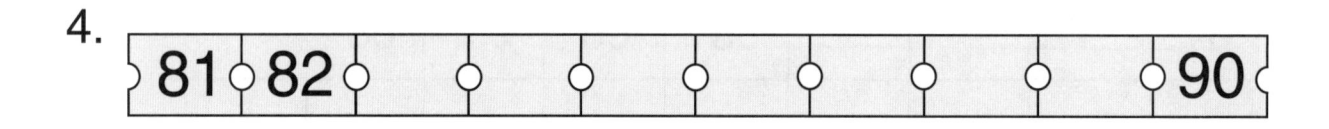

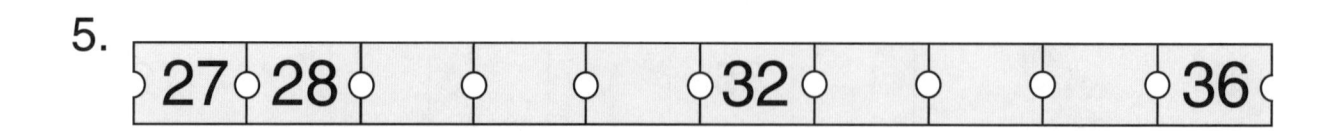

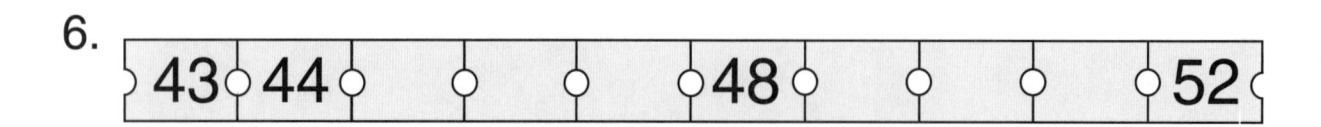

1. 21 22 __ __ __ __ __ __ __ 30

2. 31 32 __ __ __ __ __ __ __ 40

3. 51 52 __ __ __ __ __ __ __ 60

4. 81 82 __ __ __ __ __ __ __ 90

5. 27 28 __ __ __ 32 __ __ __ 36

6. 43 44 __ __ __ 48 __ __ __ 52

138
© Math Teachers Press, Inc. Reproduction by any means is strictly prohibited.

Name _____

Count by 10s to 100.

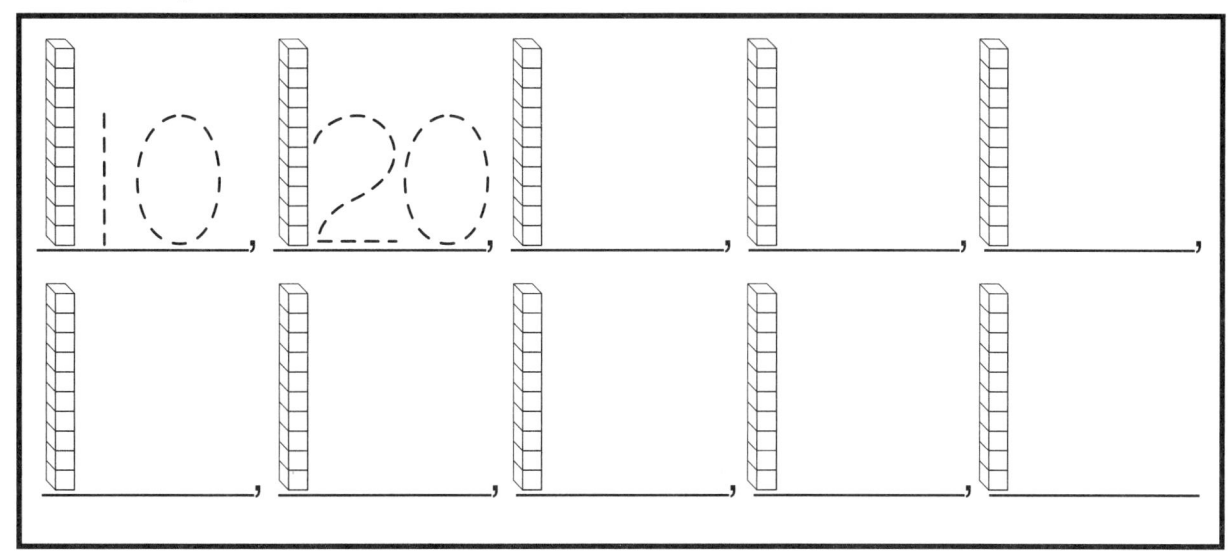

10, 20, ___, ___, ___,
___, ___, ___, ___, ___

1. Write the numeral.

5 tens _____

8 tens _____

2 tens _____

9 tens _____

7 tens _____

1 ten _____

6 tens _____

10 tens _____

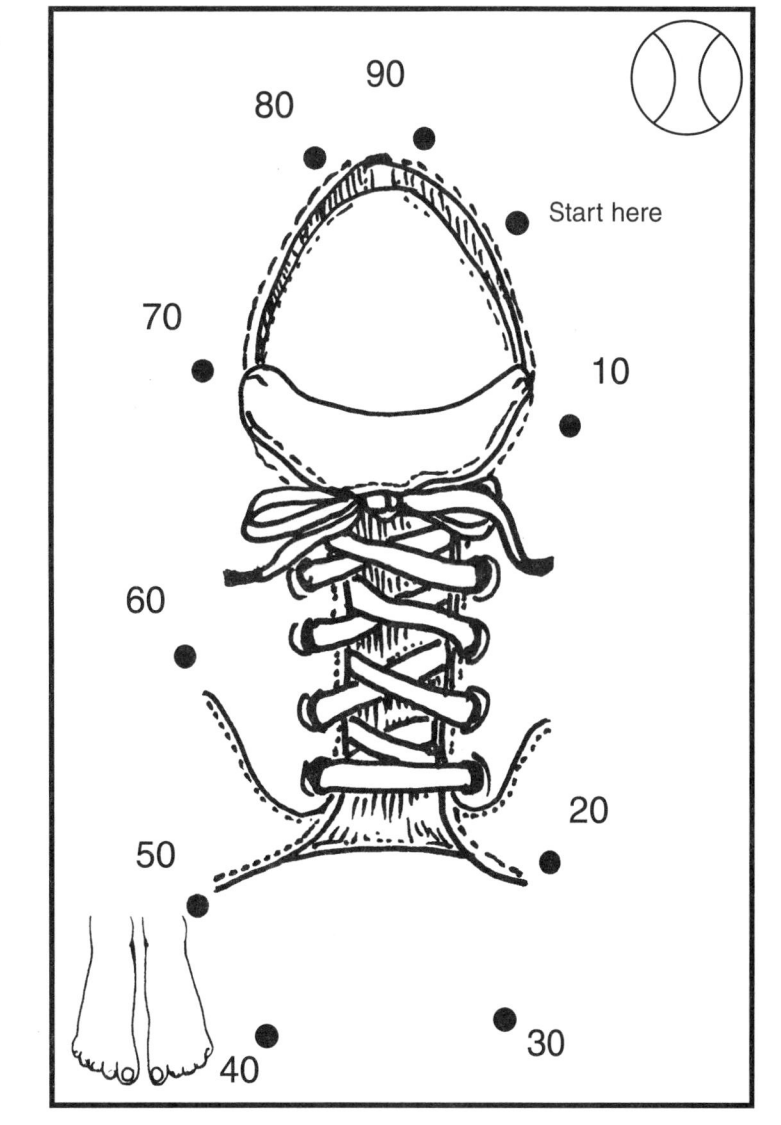

Name _____

Counting by Tens to 100

Build. Count by tens. Trace each number and word.

10 ten	20 twenty
30 thirty	40 forty
50 fifty	60 sixty
70 seventy	80 eighty
90 ninety	100 one hundred

Name _____

How Many Ways to 100?

1. Count by 2s from 2 to 100. Color each landing point.
2. Count by 5s to 100. Ring each landing point.
3. Count by 10s to 100. Draw an X on each landing point.

1	2	3	4	5	6	7	8	9	10
11	12	13	14	15	16	17	18	19	20
21	22	23	24	25	26	27	28	29	30
31	32	33	34	35	36	37	38	39	40
41	42	43	44	45	46	47	48	49	50
51	52	53	54	55	56	57	58	59	60
61	62	63	64	65	66	67	68	69	70
71	72	73	74	75	76	77	78	79	80
81	82	83	84	85	86	87	88	89	90
91	92	93	94	95	96	97	98	99	100

4. Fill in the blanks.

5, 10, 15, ____, ____, ____, ____, ____, ____, 50,

55, ____, ____, ____, ____, ____, ____, ____, ____, 100

Name _____

Complete the Patterns

1. 2, 4, ____, ____, ____, ____, ____, ____, ____, 20
2. 5, 10, ____, ____, ____, ____, ____, ____, ____, 50
3. 36, 38, ____, ____, ____, ____, ____, ____, 52
4. 25, 30, 35, ____, ____, ____, ____, ____, 65
5. Count by 2s. Color.

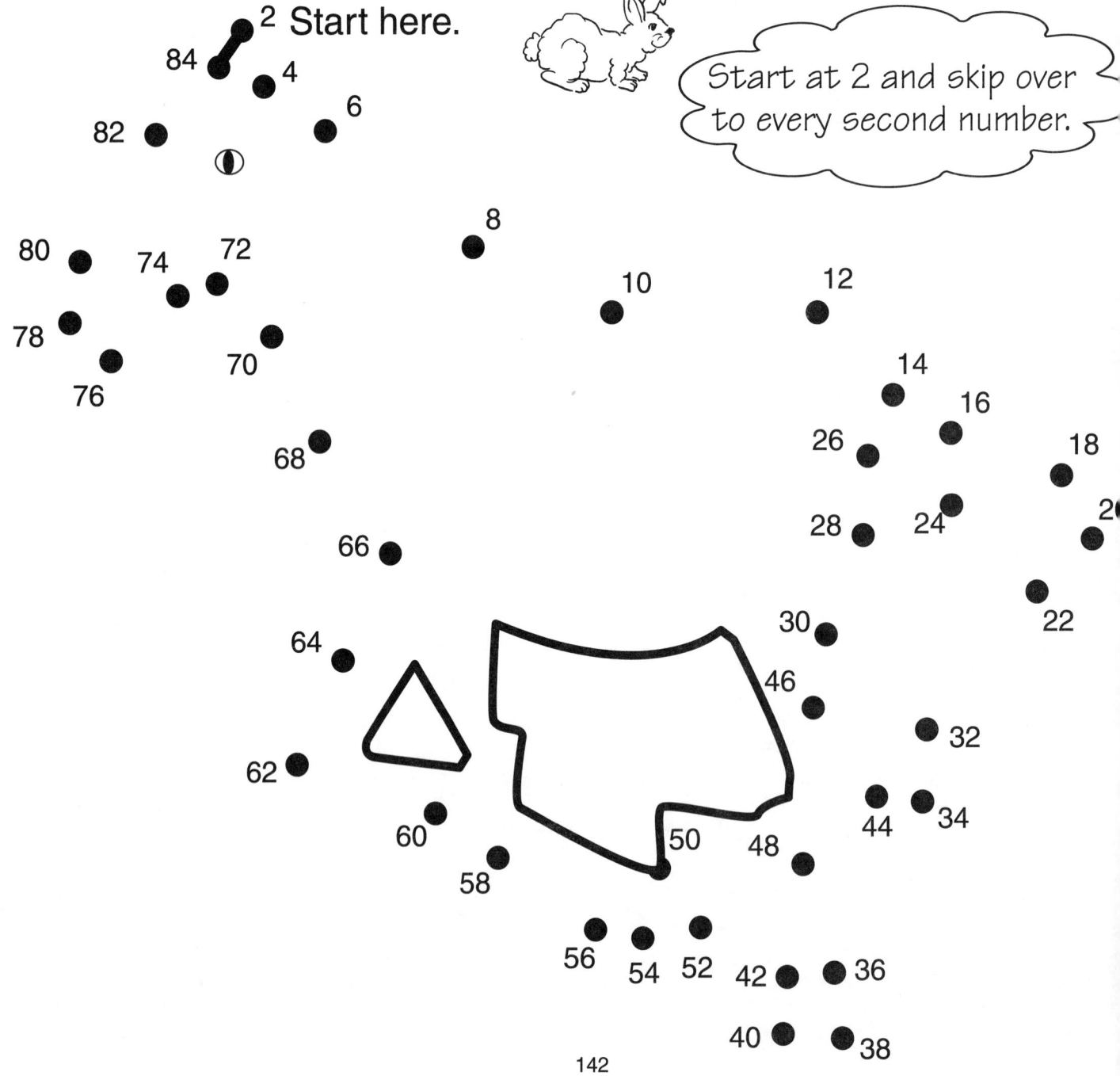

Start at 2 and skip over to every second number.

Name _____

How Many Tens and Ones?

The Toasty Cereal box has 27 cubes.
How many sticks of 10?
How many left over?

27 = ____ sticks of 10 ____ left over = ____ tens ____ ones

Ring sets of 10. How many groups of 10? How many left over? How many tens? How many ones? How many in all?

1.   = ____ groups of 10 ____ left over
     ____ tens ____ ones = ____ in all

2.   = ____ groups of 10 ____ left over
     ____ tens ____ ones = ____ in all

3.   = ____ groups of 10 ____ left over
     ____ tens ____ ones = ____ in all

4.   = ____ groups of 10 ____ left over
     ____ tens ____ ones = ____ in all

Name _____

How many tens and ones? Write the numerals.

1. | Tens | Ones |
|------|------|
| | |

2. | Tens | Ones |
|------|------|
| | |

3. | Tens | Ones |
|------|------|
| | |

4. | Tens | Ones |
|------|------|
| | |

5. | Tens | Ones |
|------|------|
| | |

6. | Tens | Ones |
|------|------|
| | |

7. | Tens | Ones |
|------|------|
| | |

8. | Tens | Ones |
|------|------|
| | |

9. | Tens | Ones |
|------|------|
| | |

10. | Tens | Ones |
|------|------|
| | |

11. | Tens | Ones |
|------|------|
| | |

12. | Tens | Ones |
|------|------|
| | |

Name _____

Build sticks of ten and ones to match each number.
Shade a picture of each number.

1. 32 = _____ tens _____ ones

Color _____ strips and _____ ones

2. 46 = _____ tens _____ ones

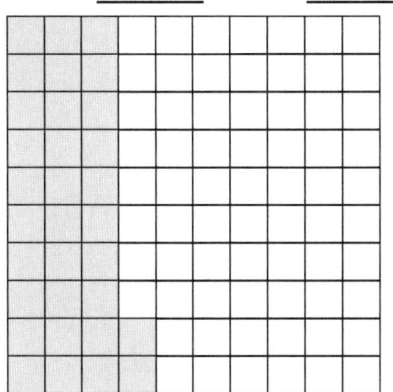

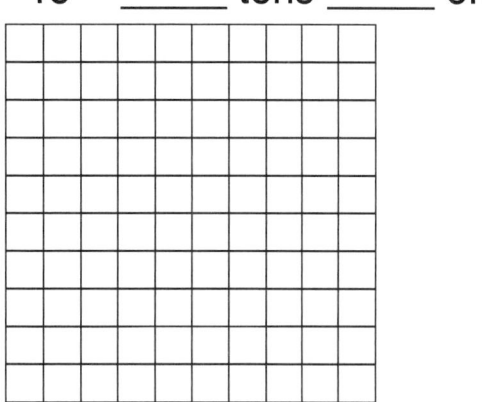

Color _____ strips and _____ ones

3. 72 = _____ tens _____ ones

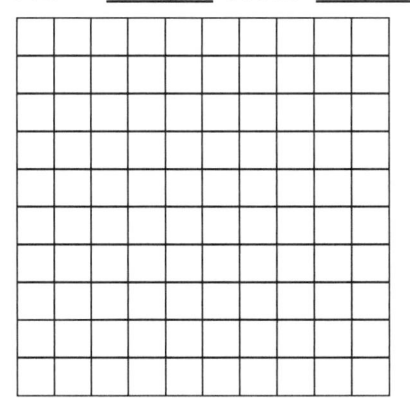

Color _____ strips and _____ ones

4. 87 = _____ tens _____ ones

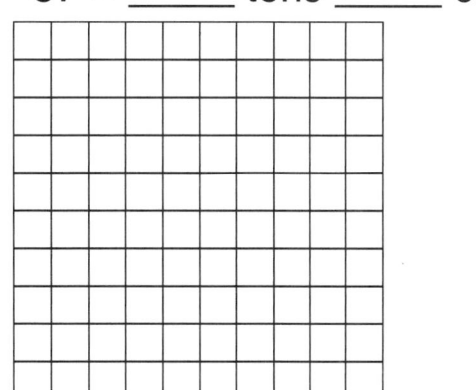

Color _____ strips and _____ ones

Draw a picture of each number.

5. 63 = |||||| :

6. 54 =

7. 27 =

8. 30 =

9. 14 =

10. 94 =

Name _____

Find the number. Shade the bubble under the number.

1. Which number has 8 in the tens place?

   | 68 | 86 |
   | ○ | ○ |

2. Which number has 3 in the tens place?

   | 31 | 13 |
   | ○ | ○ |

3. Which number has 4 in the tens place?

   | 14 | 41 |
   | ○ | ○ |

4. Which number has 7 in the tens place?

   | 17 | 71 |
   | ○ | ○ |

5. Which number has 6 in the ones place?

   | 63 | 36 |
   | ○ | ○ |

6. Which number has 2 in the ones place?

   | 21 | 12 |
   | ○ | ○ |

7. Which number has 9 in the ones place?

   | 96 | 69 |
   | ○ | ○ |

8. Which number has 1 in the tens place?

   | 81 | 18 |
   | ○ | ○ |

9. How many? _____

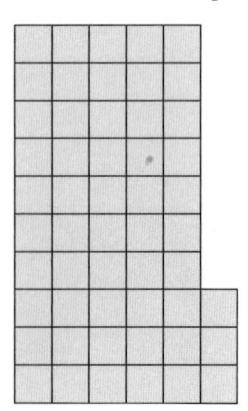

10. How many? _____

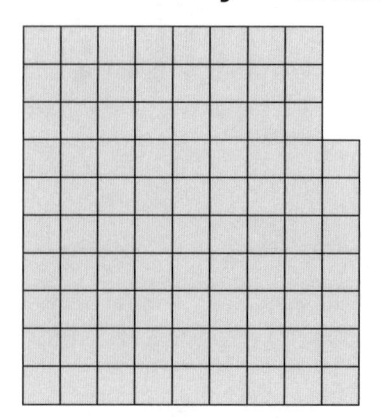

Name _____

Comparing Numbers

Fill with cubes.
Which holds more? | Which holds less?

___ tens ___ ones ___ tens ___ ones | ___ tens ___ ones ___ tens ___ ones
_____ _____ | _____ _____

1. Ring the greater number. | 2. Ring the number that is less.

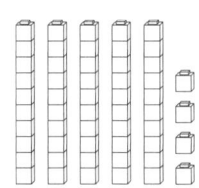

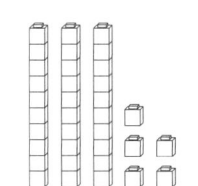

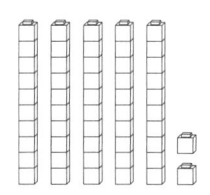

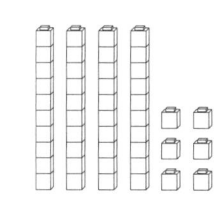

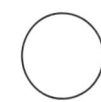

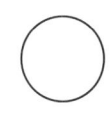

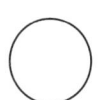

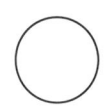

3.
|||| ⁖   is less than ||||
        is more than

42 (>) 40

4.
||||| ⁖   is less than |||||
         is more than

○ _____ _____

<, >, = ?

5.   42 ○ 48 | 6.   53 ○ 46

7.   29 ○ 30 | 8.   74 ○ 72

147
© Math Teachers Press, Inc. Reproduction by any means is strictly prohibited.

Name _____

Write the number. >, < ?

1. 23 < 26

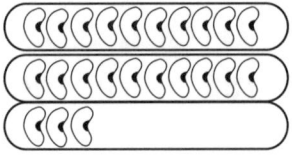

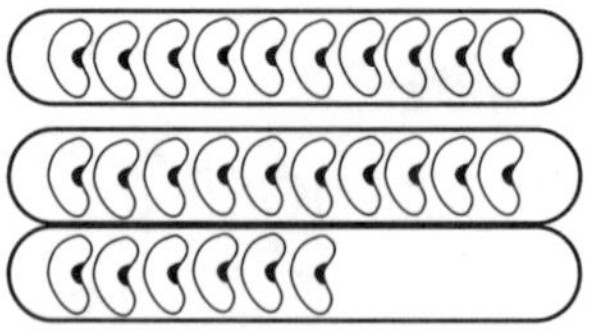

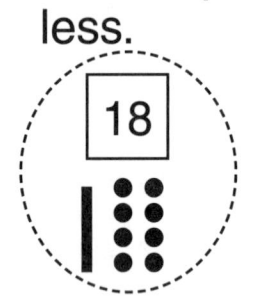

2. ___ ○ ___

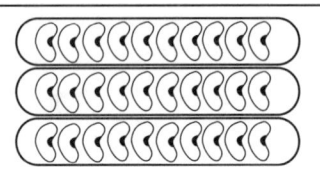

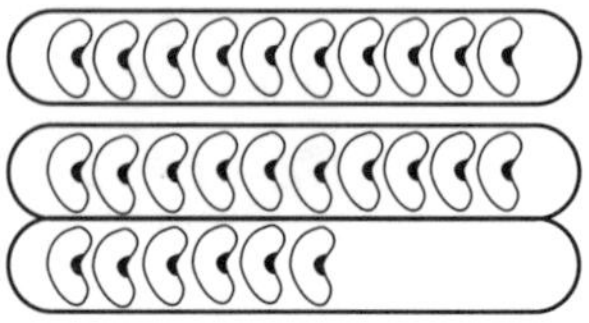

3. Draw a picture. Ring less.

   18   81

4. Draw a picture. Ring the greater number.

   74   47

5. Which is greater?

   35   53
   ○   ○

6. Which is greater?

   76   67
   ○   ○

7. Which is less?

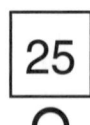

   25   52
   ○   ○

8. Which is greater?

   37   73
   ○   ○

8. Which is greater?

   87   78
   ○   ○

10. Which is less?

    46   64
    ○   ○

Name _____

Chapter 5 Review

1. How many bears? Ring the number.

   12   13   14

2. Count. Ring groups of 10. Write how many.

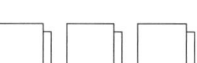

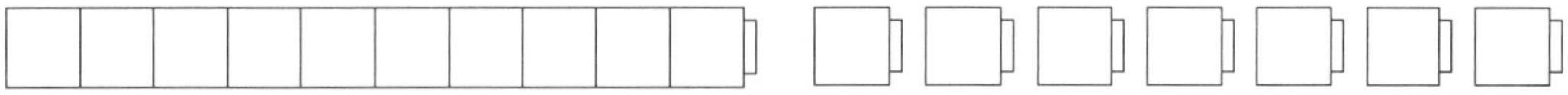

   _____ tens _____ ones =

3. Which number is less? Ring less.

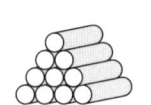

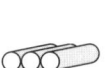

   13

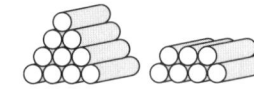

   17

4. Ring the greater number.

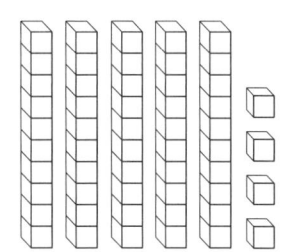

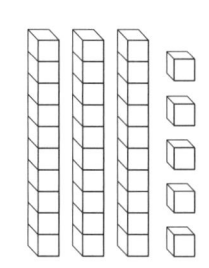

5. Count by tens. Write how many.

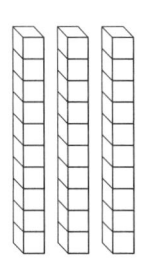

   _____

6. Count. How many?

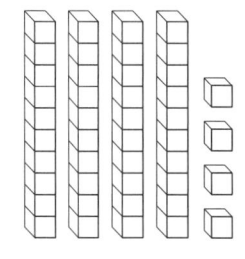

   _____ tens _____ ones

   _____

Name _____

7. What number is missing?
   Ring the missing number.

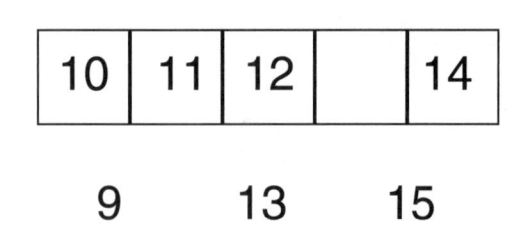

   | 10 | 11 | 12 | | 14 |

   9   13   15

8. What numbers come before 17?
   Write the missing numbers.

   _____, _____, _____, 17

9. Complete the pattern. Write the missing numbers.

   5, 10, _____, _____, _____, 30

10. Complete the pattern. Write the missing numbers.

    34, 36, _____, _____, 42

11. Which number has 3 in the tens place?
    Shade the bubble.

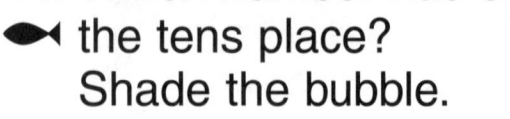

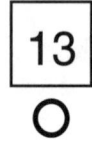

    | 31 | | 13 |
     ○   ○

12. Which number has 9 in the ones place?
    Shade the bubble.

    | 19 | | 91 |
     ○   ○

150
© Math Teachers Press, Inc. Reproduction by any means is strictly prohibited.

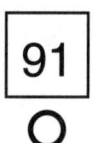

6
Time, Money, Measurement

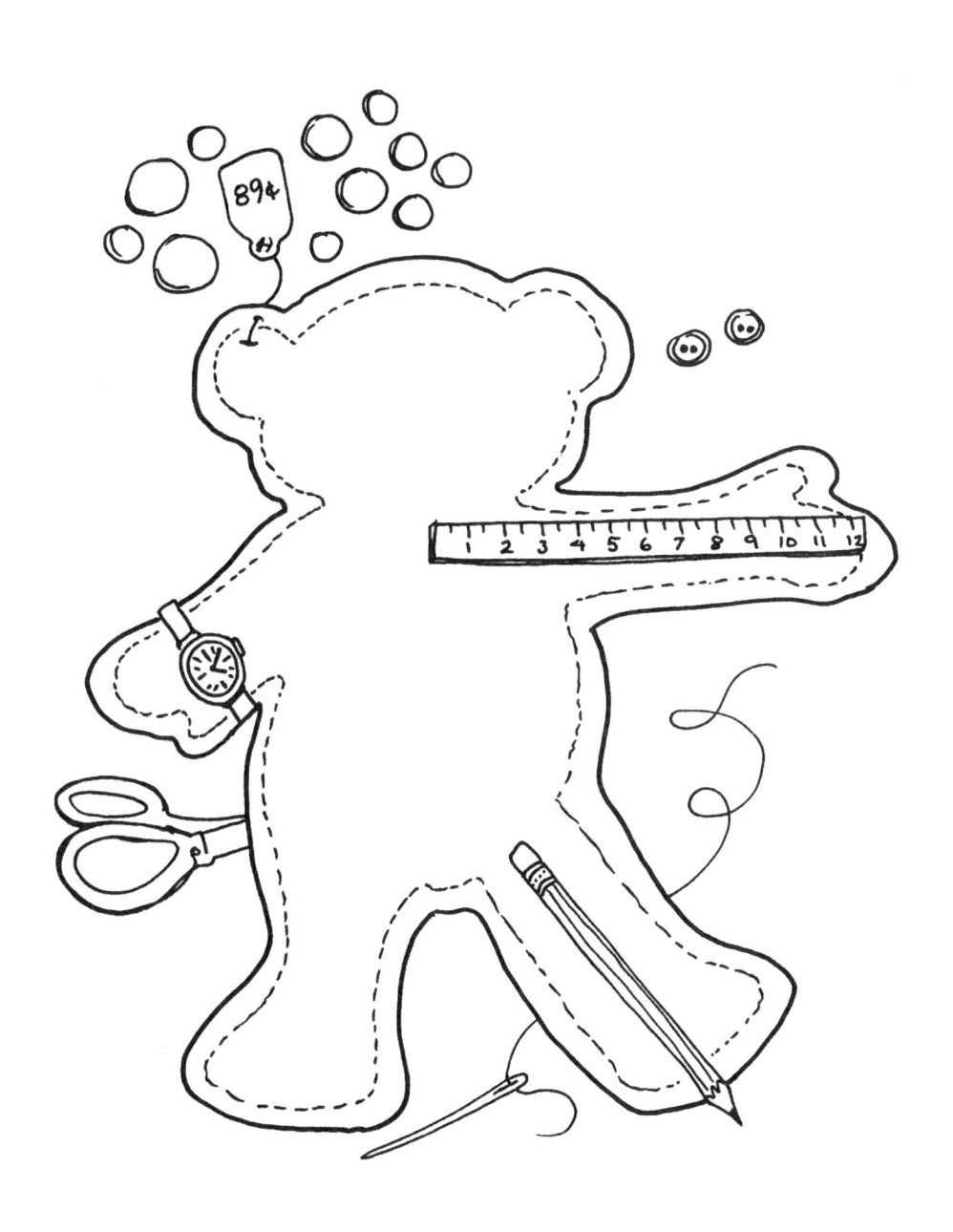

Name _____

Calendar

_____

Sunday	Monday	Tuesday	Wednesday	Thursday	Friday	Saturday
			1	2	3	4
5	6	7	8	9	10	11
12	13	14	15	16	17	18
19	20	21	22	23	24	25
26	27	28	29	30	31	

1. This month comes after December. Write the name of the month on the calendar.

2. What day of the week is January 15?
   Sunday ○   Monday ○   Tuesday ○   Wednesday ○

3. How many days are in this month?
   28 ○   29 ○   30 ○   31 ○

4. How many Fridays are in this month?
   3 ○   4 ○   5 ○   6 ○

5. The third Tuesday falls on what date of the month?
   7 ○   14 ○   21 ○   28 ○

152
© Math Teachers Press, Inc. Reproduction by any means is strictly prohibited.

Name _____

hour hand __2__

minute hand __12__

_____ o'clock

2:00

The minute hand points to 12 when it is o'clock.

Color the hour hand red. Color the minute hand blue.
What is the time?

1. hour _____
minute _____
_____ o'clock
_____

2. hour _____
minute _____
_____ o'clock
_____

3. hour _____
minute _____
_____ o'clock
_____

4. hour _____
minute _____
_____ o'clock
_____

5. hour _____
minute _____
_____ o'clock
_____

6. hour _____
minute _____
_____ o'clock
_____

Name _____

Elapsed Time

Pedro went to play basketball at 4:00.
He played for 1 hour.
What time did he stop playing?

The clock starts at 4 o'clock. 1 more hour passes. 4 hours plus 1 more hour is 5 hours.

4:00   _____

---

1. Pat went biking at 3:00.
   She biked for 1 hour.
   What time did she finish?

   2 o'clock   4 o'clock   5 o'clock
   ○   ○   ○

2. Don went shopping with his mother at 2:00.
   He got home 1 hour later.
   What time did he get home?

   1 o'clock   3 o'clock   4 o'clock
   ○   ○   ○

3. Meg went fishing at 9:00
   She fished for 2 hours.
   What time did she finish?

   10 o'clock 11 o'clock 12 o'clock
   ○   ○   ○

4. Kim went to a movie at 6:00.
   The movie lasted 2 hours.
   What time did the movie end?

   5 o'clock   7 o'clock   8 o'clock
   ○   ○   ○

5. Dick began reading at 9:00.
   He read for 1 hour.
   What time did he finish?
   Draw the times on the clocks.

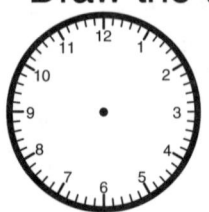

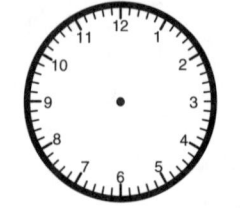

   Begin: _____   Finish: _____

6. Jan began playing at 3:00.
   She played for 2 hours.
   What time did she finish?
   Draw the times on the clocks.

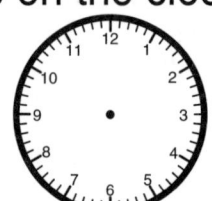

   Begin: _____   Finish: _____

Name _____

Telling Time to the Half Hour

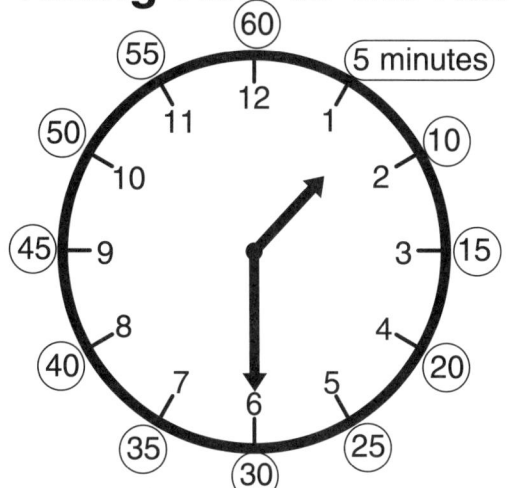

The hour hand points between ____ and ____.

It is after ____ o'clock.

The minute hand points to ____.

It is ____ minutes after 1 o'clock.

"one thirty"   1:30

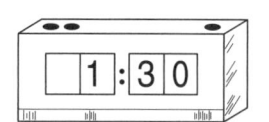

Ring the correct time.

1. 6:30
   7:00
   7:30

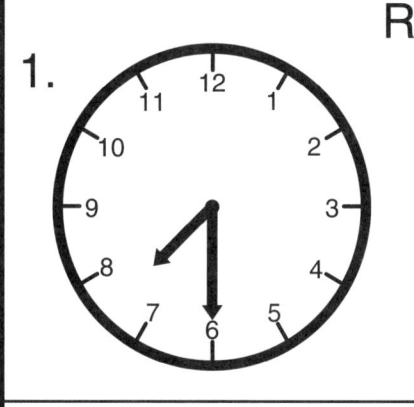

2. 2:30
   3:00
   3:30

3. 5:30
   6:00
   6:30

4. 9:30
   10:00
   10:30

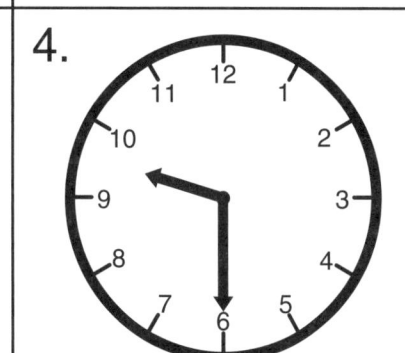

5. Draw hands to show 3:30.

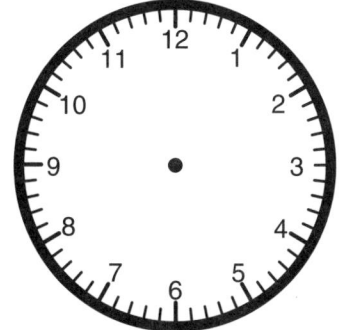

6. Draw hands to show 8:30.

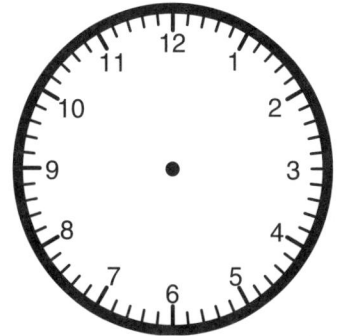

Name _____

Connect the times which match.

1.
2.
3.
4.
5.
6.

11:00

3:00

7:00

4:30

1:30

10:30

Name _____

A Fair Trade: 1 Nickel for 5 Pennies

> 1 nickel has the same value as 5 pennies. 
>
> 5¢ = 5¢
>
> *This is a fair trade.*

Count by nickels. How much money?

1. *5, 10, 10¢* _____ ¢

2. _____ ¢

3. _____ ¢

4. _____ ¢

Count on.

5. 5¢ 6¢ 7¢ 8¢ _____ ¢

6. _____ ¢

157
© Math Teachers Press, Inc. Reproduction by any means is strictly prohibited.

Name _____

Problem Solving: Acting Out the Problem

1. Kim had [5¢ nickel] [1¢ penny]

    She bought [notepad 8¢]

    How much is left? _____ ¢

2. Pedro had [1¢ penny] [1¢ penny]

    He bought [scissors 5¢]

    How much is left? _____ ¢

3. Meg had [1¢ penny] [5¢ nickel]

    She spent [1¢] [1¢] [1¢]

    How much is left? _____ ¢

4. Don had [1¢] [1¢] [1¢] [1¢] [1¢]

    He lost [1¢] [1¢] [1¢]

    How much is left? _____ ¢

5. Ben had [5¢] [1¢] [1¢] [1¢] [1¢]

    He bought [pen 6¢]

    How much is left? _____ ¢

6. Pat had [5¢] [1¢] [1¢]

    She bought [ring 5¢]

    How much is left? _____ ¢

Name _____

A Fair Trade for 1 Dime

1 dime has the same value as 10 pennies.

10¢ = 10¢

This is a fair trade.

____ dime = ____ nickels

____ dime = ____ nickels ____ pennies

How much money?

1. _____ ¢

2. _____ ¢

3. _____ ¢

4. _____ ¢

5. _____ ¢

6. _____ ¢

Name _____

Count by tens, fives and ones to find how much.

10¢   20¢   ___¢   ___¢   ___¢   ___¢

1	2	3	4	5	6	7	8	9	10
11	12	13	14	15	16	17	18	19	20
21	22	23	24	25	26	27	28	29	30
31	32	33	34	35	36	37	38	39	40
41	42	43	44	45	46	47	48	49	50
51	52	53	54	55	56	57	58	59	60
61	62	63	64	65	66	67	68	69	70
71	72	73	74	75	76	77	78	79	80
81	82	83	84	85	86	87	88	89	90
91	92	93	94	95	96	97	98	99	100

Write the amount.

1. _____¢

2. _____¢

3. _____¢

4. _____¢

How could you pay?
Use the fewest number of coins.

5. apple 29¢

6. banana 73¢

7. grapes 38¢

160
© Math Teachers Press, Inc. Reproduction by any means is strictly prohibited.

Name _____

A Fair Trade for 1 Quarter

This is a fair trade.

1 quarter has the same value as 25 pennies. 25¢

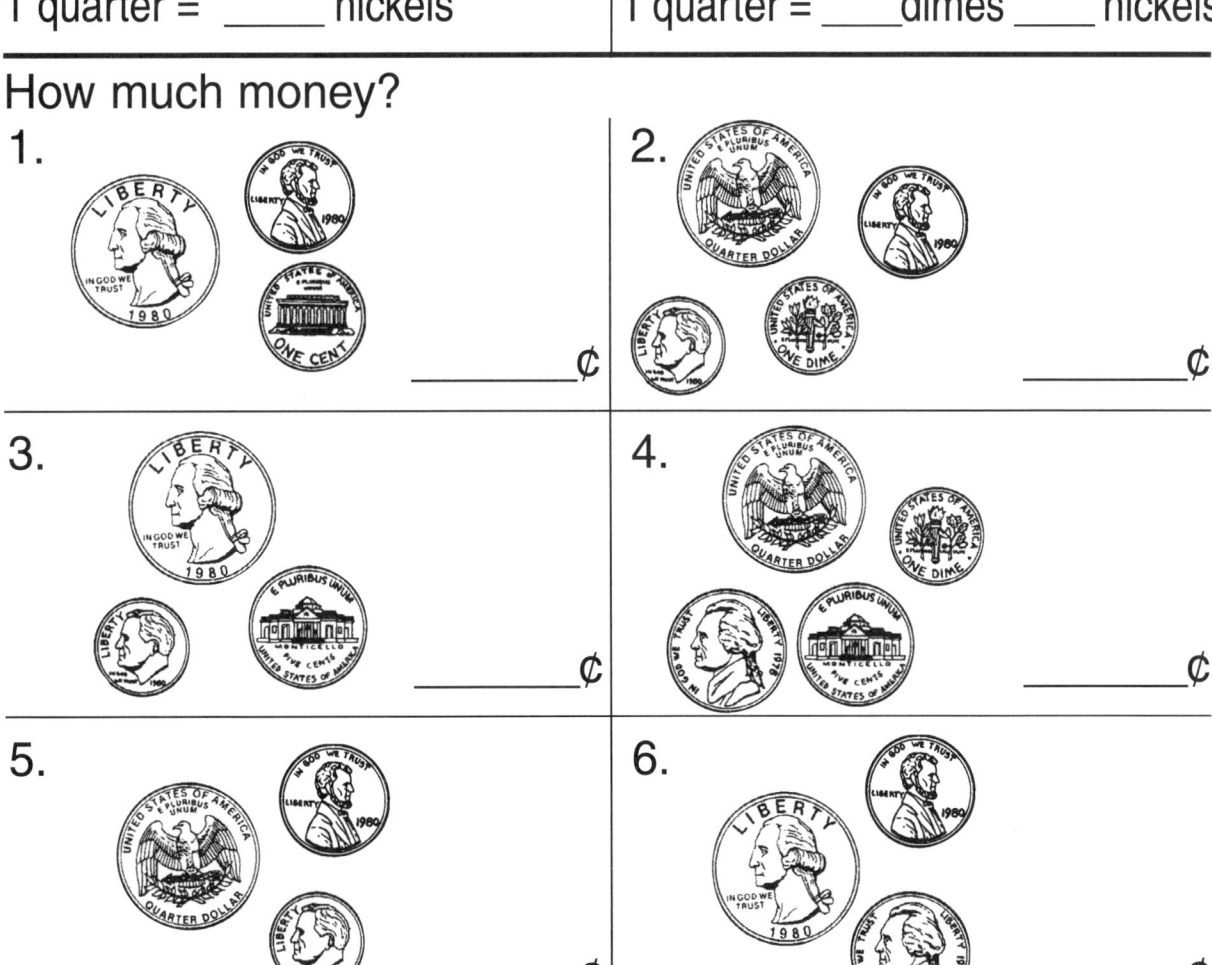

1 quarter = _____ nickels

1 quarter = _____ dimes _____ nickels

How much money?

1. _____ ¢

2. _____ ¢

3. _____ ¢

4. _____ ¢

5. _____ ¢

6. _____ ¢

Name _____

Is There Enough Money to Buy?

Start with 25. Then count by tens, fives and ones to find the amount.

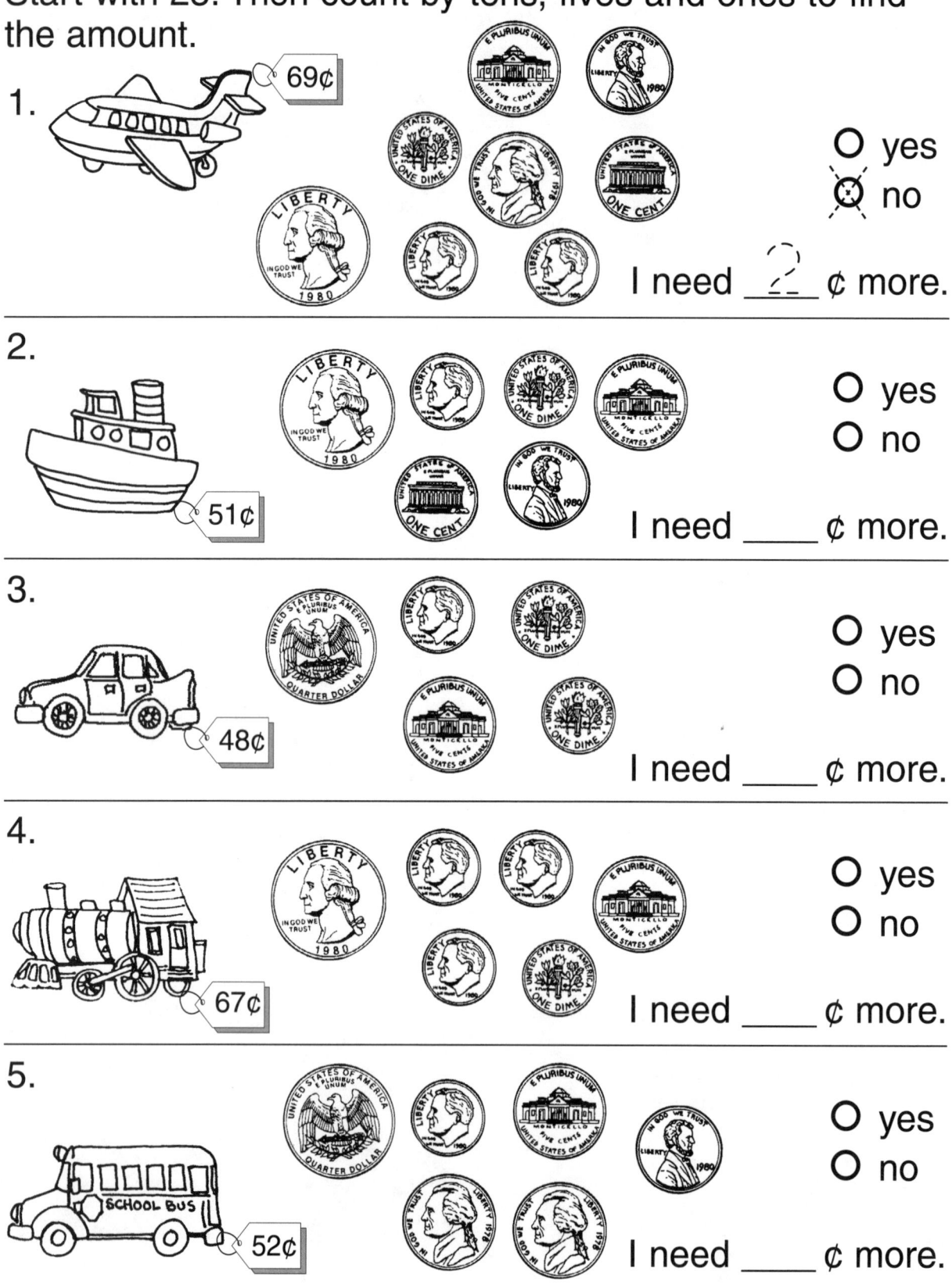

1. Airplane — 69¢
   ○ yes
   ⊗ no
   I need __2__ ¢ more.

2. Boat — 51¢
   ○ yes
   ○ no
   I need ____ ¢ more.

3. Car — 48¢
   ○ yes
   ○ no
   I need ____ ¢ more.

4. Train — 67¢
   ○ yes
   ○ no
   I need ____ ¢ more.

5. School Bus — 52¢
   ○ yes
   ○ no
   I need ____ ¢ more.

Name _____

1. Which is more? Ring more.

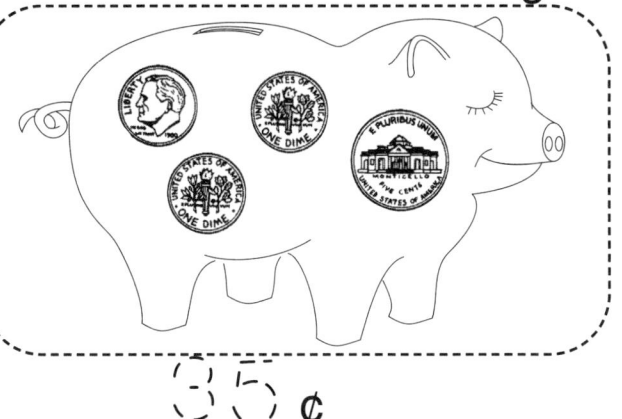

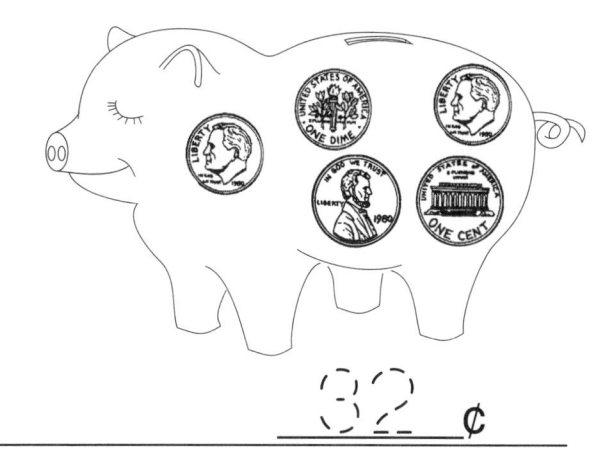

_____35_____ ¢   _____32_____ ¢

2. Which is more? Ring more.

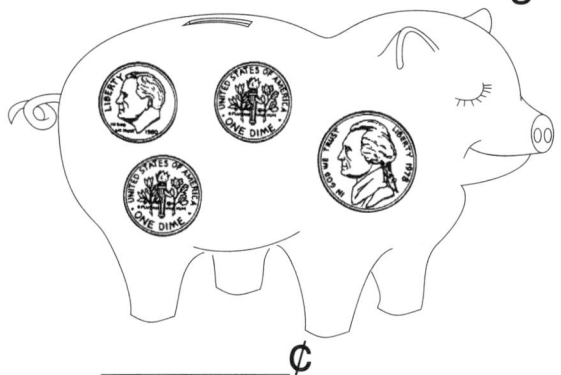

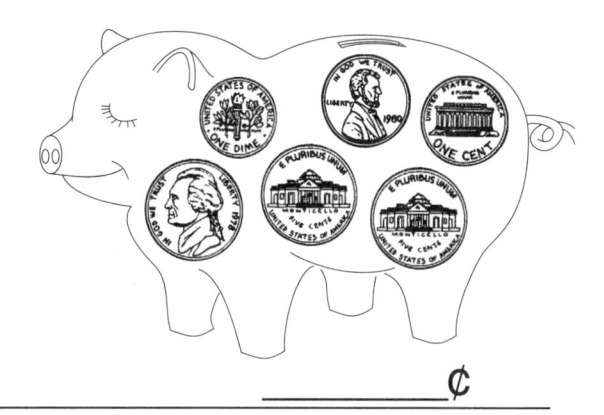

_____ ¢   _____ ¢

3. Which is less? Ring less.

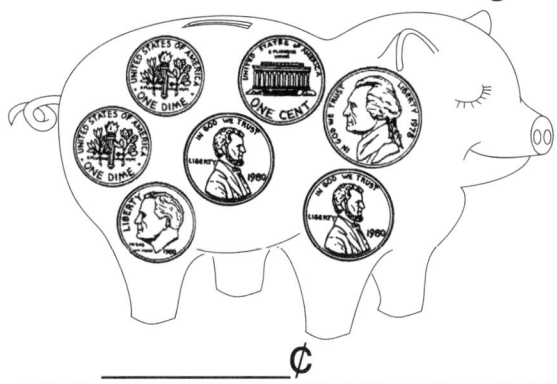

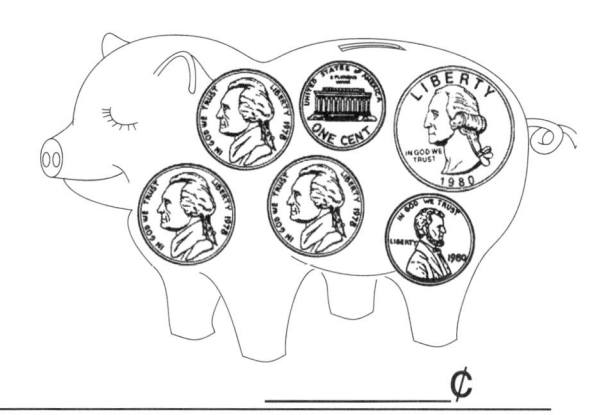

_____ ¢   _____ ¢

4. Which is less? Ring less.

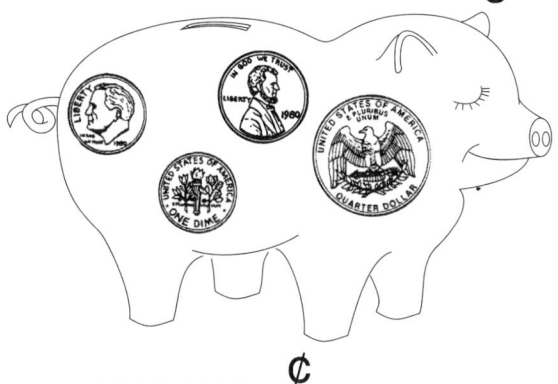

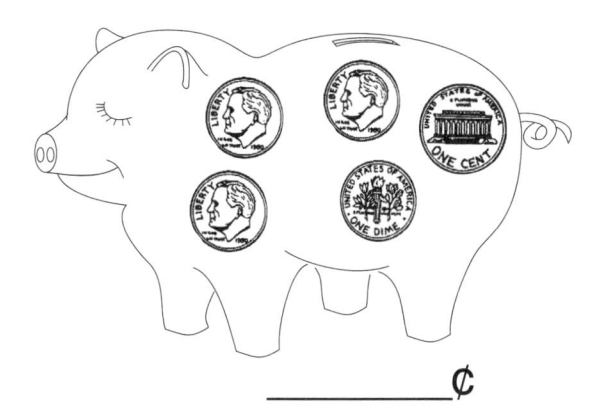

_____ ¢   _____ ¢

Name _____

Patterns
Some patterns repeat. Some do not repeat.
Guess what comes next in these patterns.

P Q R S P Q R S ___ ___ ___ ___

20, 25, 30, 35, ___, ___, ___, 55

Look for the pattern. Write the missing numbers or letters.

1. 21 22 23 24 21 22 23 24 ___ ___ ___ ___

2. A B B A B B ___ ___ ___

3. 22 24 26 28 ___ ___ ___ 28

4. 5 10 15 20 5 ___ ___ 20

5. 5 15 25 ___ ___ ___ 65

6. 7 17 27 ___ ___ ___ 67

7. A B B A A B B A ___ ___ ___ ___

8. 4 8 12 16 ___ ___ 28

Name _____

Measuring Length with Non-standard Units
How many paper clips?

1.

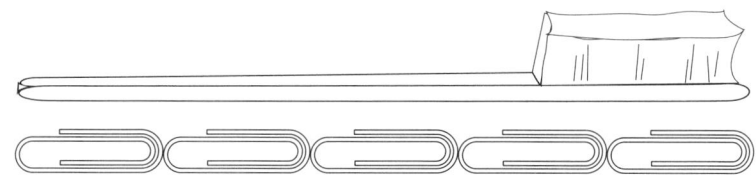

_____ units

2.

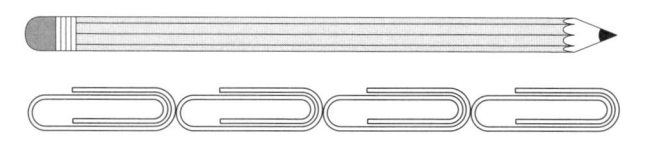

_____ units

3.

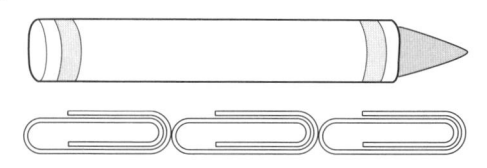

_____ units

How many cubes?

4.

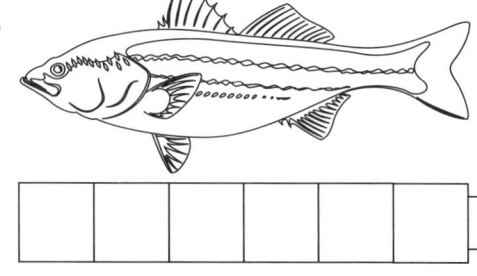

_____ cubes long

5.

6.

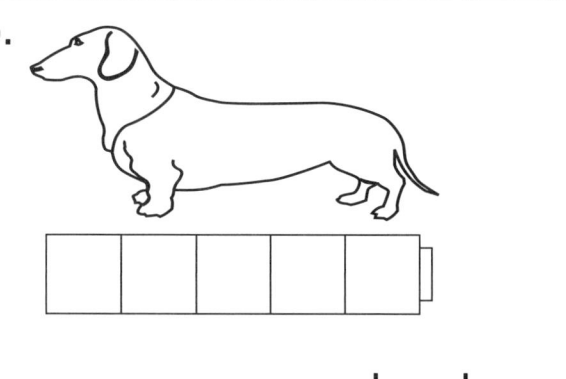

_____ cubes long

_____ cubes tall

165
© Math Teachers Press, Inc. Reproduction by any means is strictly prohibited.

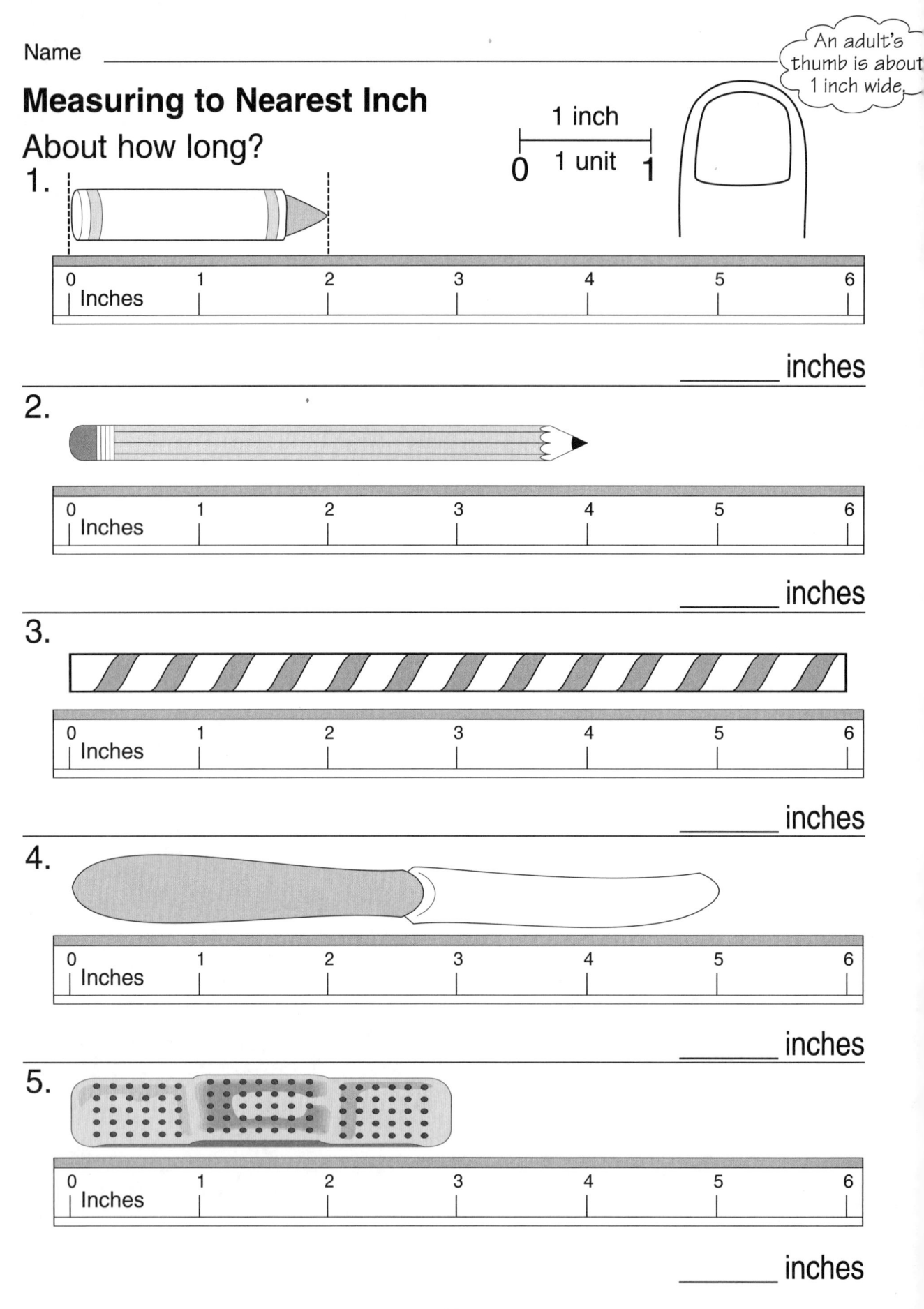

Name _____

Measuring to Nearest Inch

About how long?

An adult's thumb is about 1 inch wide.

1 inch
0 1 unit 1

1.

_____ inches

2.

_____ inches

3.

_____ inches

4.

_____ inches

5.

_____ inches

166
© Math Teachers Press, Inc. Reproduction by any means is strictly prohibited.

Name _____

Measuring to the Nearest Centimeter

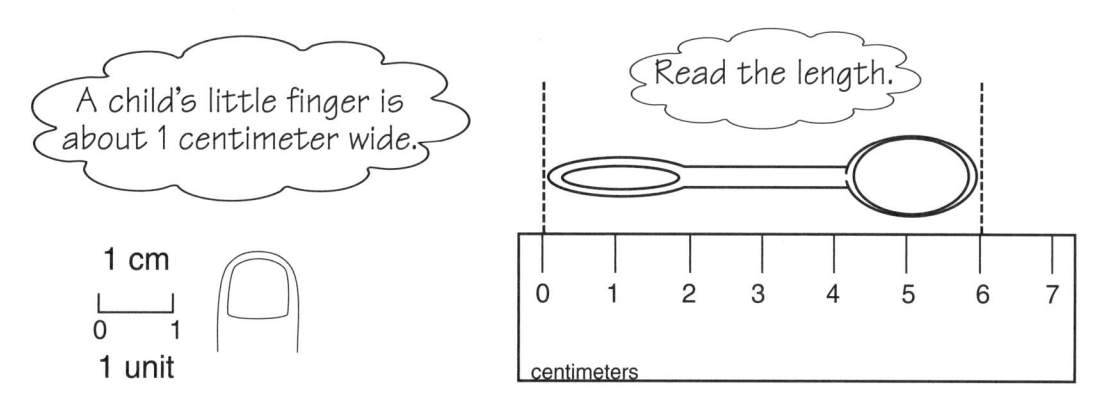

A child's little finger is about 1 centimeter wide.

1 cm
0 1
1 unit

Read the length.

0 1 2 3 4 5 6 7
centimeters

The spoon is 6 centimeters (cm) long.

About how long is each pencil?
Use blocks to measure. Check with a centimeter ruler.

1.

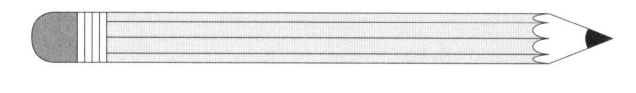

_____ cm

2.

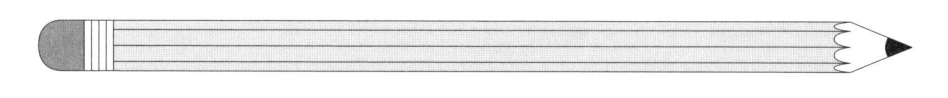

_____ cm

3.

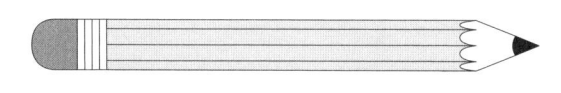

_____ cm

4.

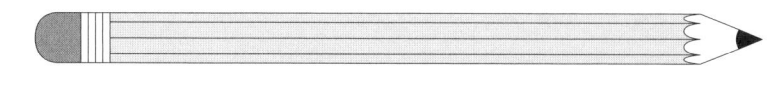

_____ cm

5.

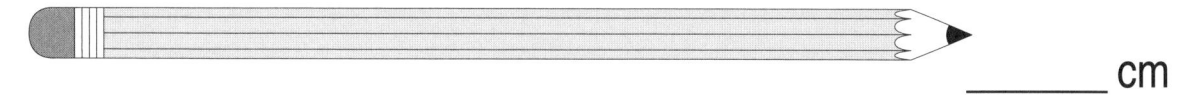

_____ cm

167
© Math Teachers Press, Inc. Reproduction by any means is strictly prohibited.

Name _____

Problem Solving: Guess and Check

How long around the track?

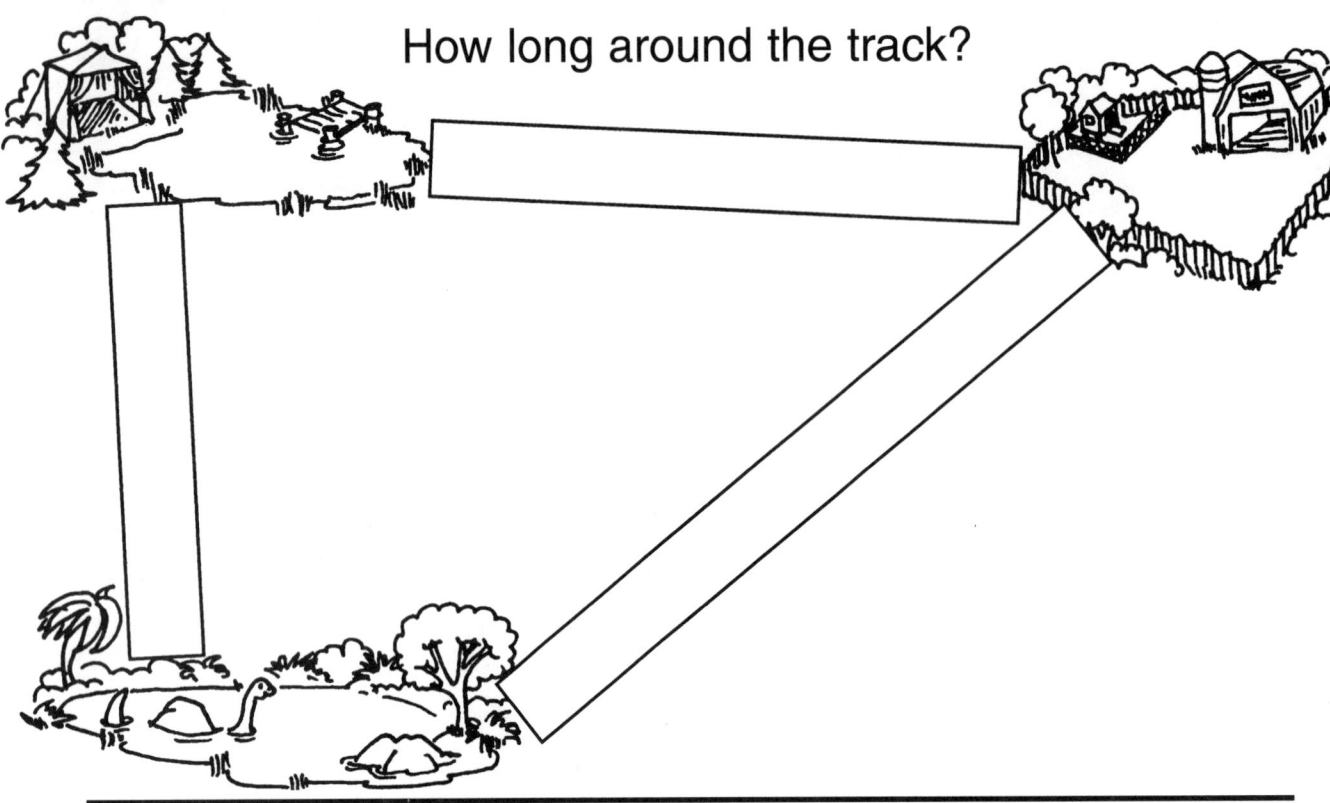

Use colored number rods. Measure the distances.

1. How far from 🏠 to 🏠 ?

   ____ white   ____ red   ____ purple   ____ brown

2. How far from 🏠 to 🌴 ?

   ____ white   ____ red   ____ yellow   ____ orange

3. How far from 🌴 to 🏠 ?

   ____ white   ____ red   ____ light green   ____ dark green

4. How far around the track?

   ____ white   ____ red

Name _____

Area

Meg is making a wall hanging with pattern block shapes. How many of each shape will cover the design?

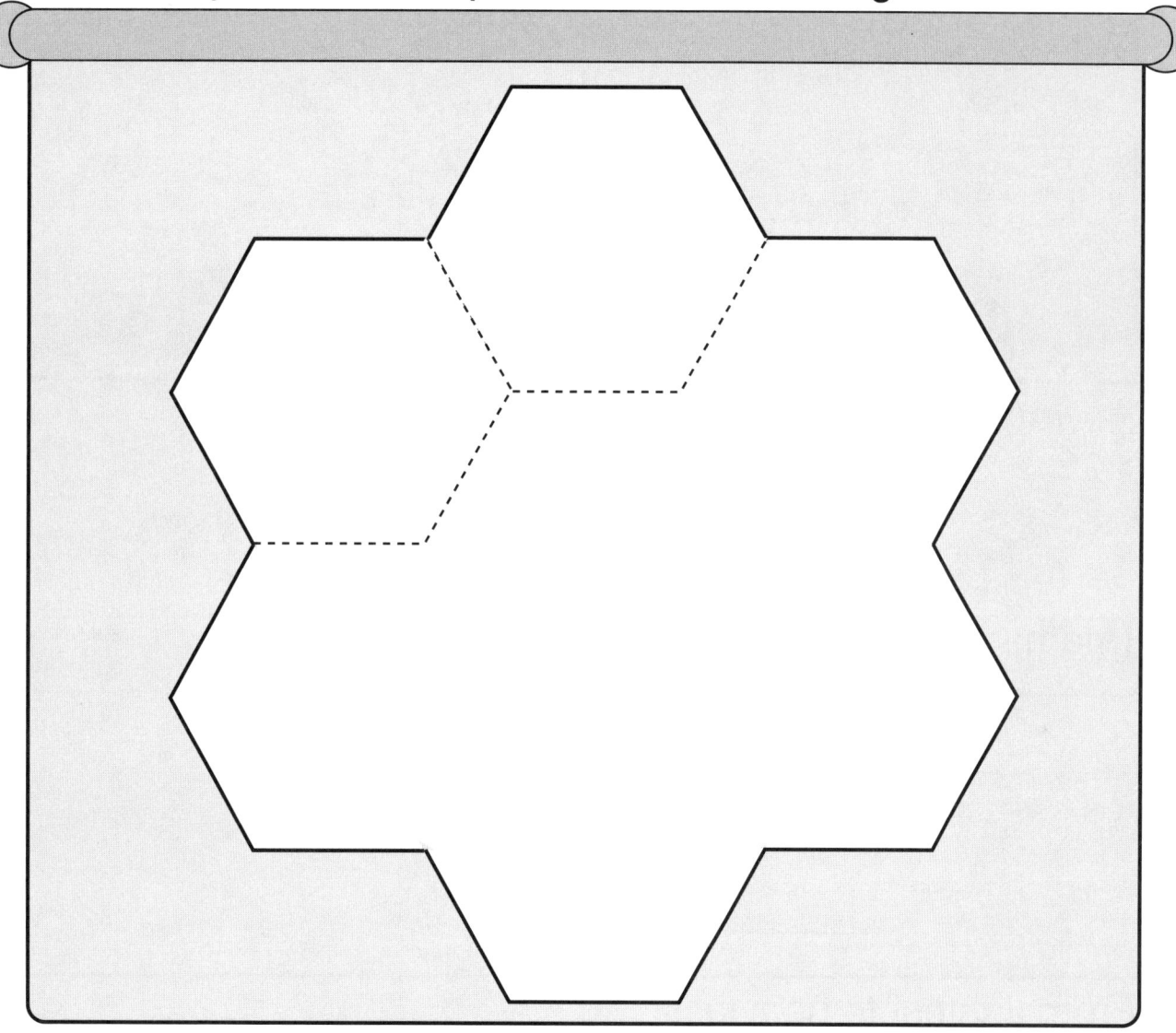

How many to cover?

1. _____ ⬡   3. _____ △

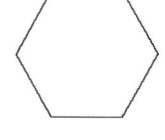

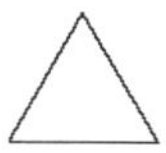

2. _____ ⏢   4. _____ ▱

5. The larger the block, the (fewer, more) blocks will be needed to cover the same space.

Name _____

Weight

| 1 papa bear weighs the same as ____ baby bears.  | 1 papa bear weighs the same as ____ paper clips.  |

How much will each object weigh in paper clips?

1.  (mama bear) ____ 📎

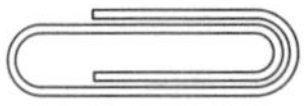

2.  (baby bear) ____ 📎

3. 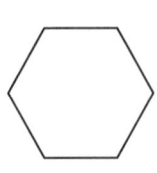 ____ 📎

4. 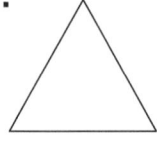 ____ 📎

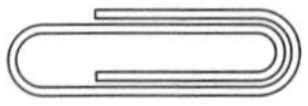

From lightest to heaviest?

5. ○ Yes ○ No

6. ○ Yes ○ No

Name _____

Capacity

2 cups fill 1 pint	4 cups fill 1 quart

Ring which holds more.

1.

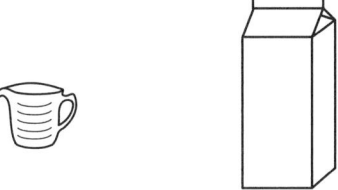

2.

3.

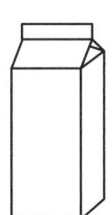

4.

Ring which holds less.

5.

6.

From holds least to holds most?

7.    ○ Yes
     ○ No

8.    ○ Yes
     ○ No

171
© Math Teachers Press, Inc. Reproduction by any means is strictly prohibited.

Name _____

Technology

Use a crayon.
Show what the LOGO turtle draws.

1. FORWARD 4
2. RIGHT 90, FORWARD 3
3. LEFT 90, FORWARD 4
4. RIGHT 90, FORWARD 4
5. RIGHT 90, FORWARD 4
6. LEFT 90, FORWARD 3
7. RIGHT 90, FORWARD 4

---

1. FORWARD 8
2. RIGHT 90, FORWARD 1
3. RIGHT 90, FORWARD 3
4. LEFT 90, FORWARD 1
5. LEFT 90, FORWARD 3
6. RIGHT 90, FORWARD 1
7. RT 90, FD 8
8. RT 90, FD 1
9. RT 90, FD 4
10. LT 90, FD 1
11. LT 90, FD 4
12. RT 90, FD 1

13. What letter is formed? _____

172

Name _____

Chapter 6 Review

1. What time is it?

   3:00   4:00   12:00
    ○   ○   ○

2. Write the time.

   ____ : ____

3. What time is it?

   2:00   2:30   6:30
    ○   ○   ○

4. Write the time.

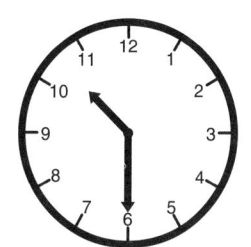

   ____ : ____

5. Write the amount.

   ____ ¢

6. How much money?

   ____ ¢

7. How much money?

   ____ ¢

173
© Math Teachers Press, Inc. Reproduction by any means is strictly prohibited.

Name _____

8. Mark the coins needed to buy the boat.

9. How many paper clips?

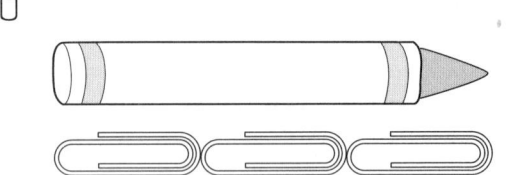

_____ paper clips

10. How many inches?

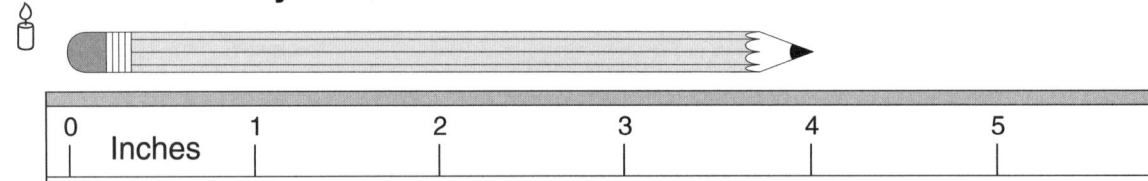

_____ inches

11. Ring which weighs more.

12. Ring which holds less.

13. Write the numbers that come next.

31, 32, 33, 34, 31, _____, _____, 34

174
© Math Teachers Press, Inc. Reproduction by any means is strictly prohibited.

7
Two-Digit Addition and Subtraction

Name _____

1. Read and understand

Each car on the train has 10 seats. The 4 Green Bears and their 10 cousins boarded the train. How many bears on the train?

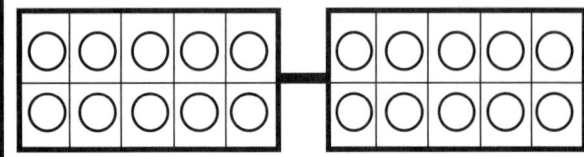

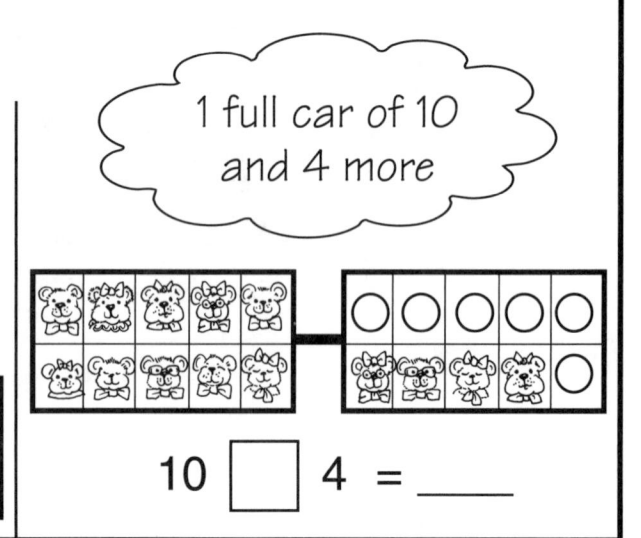

1 full car of 10 and 4 more

10 ☐ 4 = ____

1. The train with 14 bears stopped at the Children's Farm. 5 bears got on. How many bears are on the train?

   14 ⊞+ 5 = ____

2. The train with 19 bears went to Dinosaur Land. 3 bears got off. How many bears are on the train?

   19 ⊟− 3 = ____

3. The train with 16 bears stopped at Camp Wilderness. 2 bears got on the train. How many bears are on the train?

   16 ☐ 2 = ____

4. The train with 18 bears stopped at the Snack Shoppes. The 4 Green bears got off the train. How many bears are on the train?

   18 ☐ 4 = ____

5. 47 + 2 = _____   20 + 4 = _____   13 + 5 = _____

6. 21 + 6 = _____   32 + 3 = _____   45 + 4 = _____

Name _____

Using a Number Line Model

Look at the number line to help add.

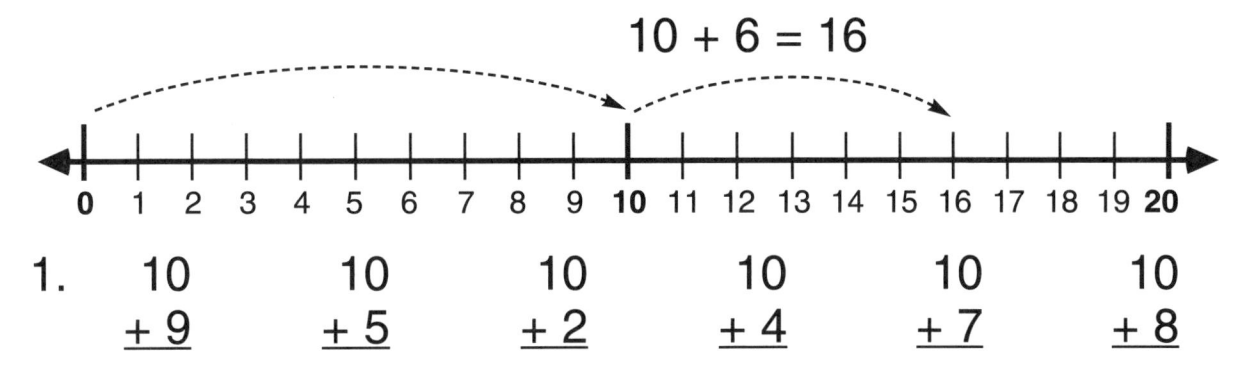

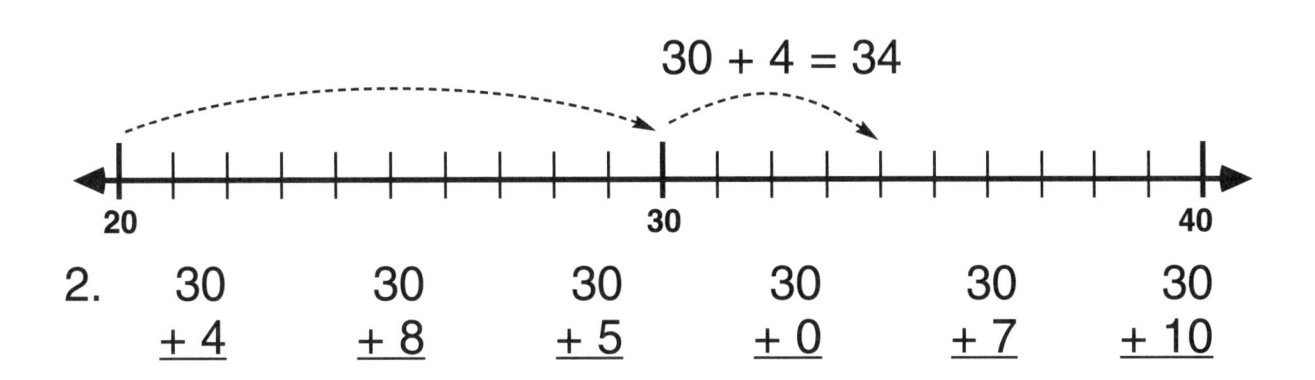

10 + 6 = 16

[number line from 0 to 20]

1. 10   10   10   10   10   10
    +9   +5   +2   +4   +7   +8

30 + 4 = 34

[number line from 20 to 40]

2. 30   30   30   30   30   30
    +4   +8   +5   +0   +7   +10

3. Pat has 10 pennies. He found 3 pennies. How many pennies does he have?

_____ pennies

4. Kim has 60 pennies. He found 5 pennies. How many pennies does he have?

_____ pennies

Name _____

Addition on the Hundred Chart

Meg bought 21 tickets for rides at the Amusement Park. Her friend gave her 4 more. How many tickets does Meg have?

1	2	3	4	5	6	7	8	9	10
11	12	13	14	15	16	17	18	19	20
21	22	23	24	25	26	27	28	29	30
31	32	33	34	35	36	37	38	39	40
41	42	43	44	45	46	47	48	49	50
51	52	53	54	55	56	57	58	59	60
61	62	63	64	65	66	67	68	69	70
71	72	73	74	75	76	77	78	79	80
81	82	83	84	85	86	87	88	89	90
91	92	93	94	95	96	97	98	99	100

Use the Hundred Chart to add.

1. Pedro had 13 tickets. Pat had 5 tickets. How many tickets did they have in all?

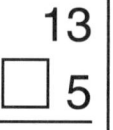

    13
    ☐ 5
    ———

2. Meg had 7 tickets. Kim had 12 tickets. How many tickets in all?

    ———

    Start with the larger number on the Hundred Chart.

3. There are 32 children on the train. There are 6 children in line. How many children are there in all?

    ———

4. There are 75 children on the train. There are 4 children in line. How many children altogether?

    ———

5.
    23
    + 4

    3
    + 32

    50
    + 2

    86
    + 3

    12
    + 5

6.
    35
    + 4

    92
    + 6

    73
    + 4

    7
    + 41

    64
    + 4

178

Name _____

> 2. Find the question and needed facts.

There are 17 bears on the train. The 4 Green Bears get off the train. How many bears are on the train?

17 ☐ 4 = _____

Underline the question. Circle the facts. Solve.

1. The train left Dinosaur Land with 16 bears on board. The train stopped at Camp Wilderness and 3 bears got off. How many bears are on the train?

    16 ☐ 3 = _____

2. The train had 18 bears on board. 4 bears got off at the Snack Shoppe. How many bears are on the train?

    18 ☐ _____ = _____

3. The bears went on an airplane trip. There were 37 bears on the plane. At the first stop, 4 bears got off. How many bears are on the plane?

    _____

4. There were 68 bears on the plane. 4 bears got off the plane. How many bears are on the plane?

    _____

5. $19 - 6 =$ _____ $25 - 5 =$ _____ $36 - 4 =$ _____

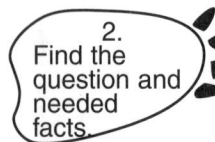

Name _____

2. Find the question and needed facts

The Green Bears bought a bag of peanuts. There were 28 peanuts in the bag. The peanuts cost 5¢. The bears ate 4 peanuts. How many peanuts are in the bag?

Cross out the peanuts eaten.

28 ☐ ___ = ___

Underline the question. Circle the facts.
Cross out facts **not** needed. Solve.

1. The Red Bears bought a medium bag of peanuts for 10¢. There were 46 peanuts in the bag. The bears ate 4 peanuts. How many peanuts are in the bag?

   ___ ☐ ___ = ___

2. The Yellow Bears bought a large bag of peanuts and a bag of chips. There were 78 peanuts in the bag. There were 46 chips in the bag. The Red Bears ate 5 chips. How many chips are left?

   ___ ☐ ___ = ___

3. The Green Bears bought 14 cartons of juice and 3 bags of peanuts. The Green Bears drank 2 cartons of juice. How many juice cartons are left?

   ___ ☐ ___ = ___

4. The Yellow Bears bought 28 juice cartons. Each carton costs 25¢. The Yellow Bears drank 4 juice cartons. How many juice cartons are left?

   ___ ☐ ___ = ___

Name _____

3. Decide on a process.

At Camp Wilderness, children make pictures by pressing leaves on paper. Pedro found 4 brown leaves and 5 green leaves. How many leaves did Pedro find?

part + part = whole
4 + 5 = ___

Pedro had 9 leaves. He used 3 leaves to make a picture. How many leaves does Pedro have?

whole − part = part
9 − 3 = ___

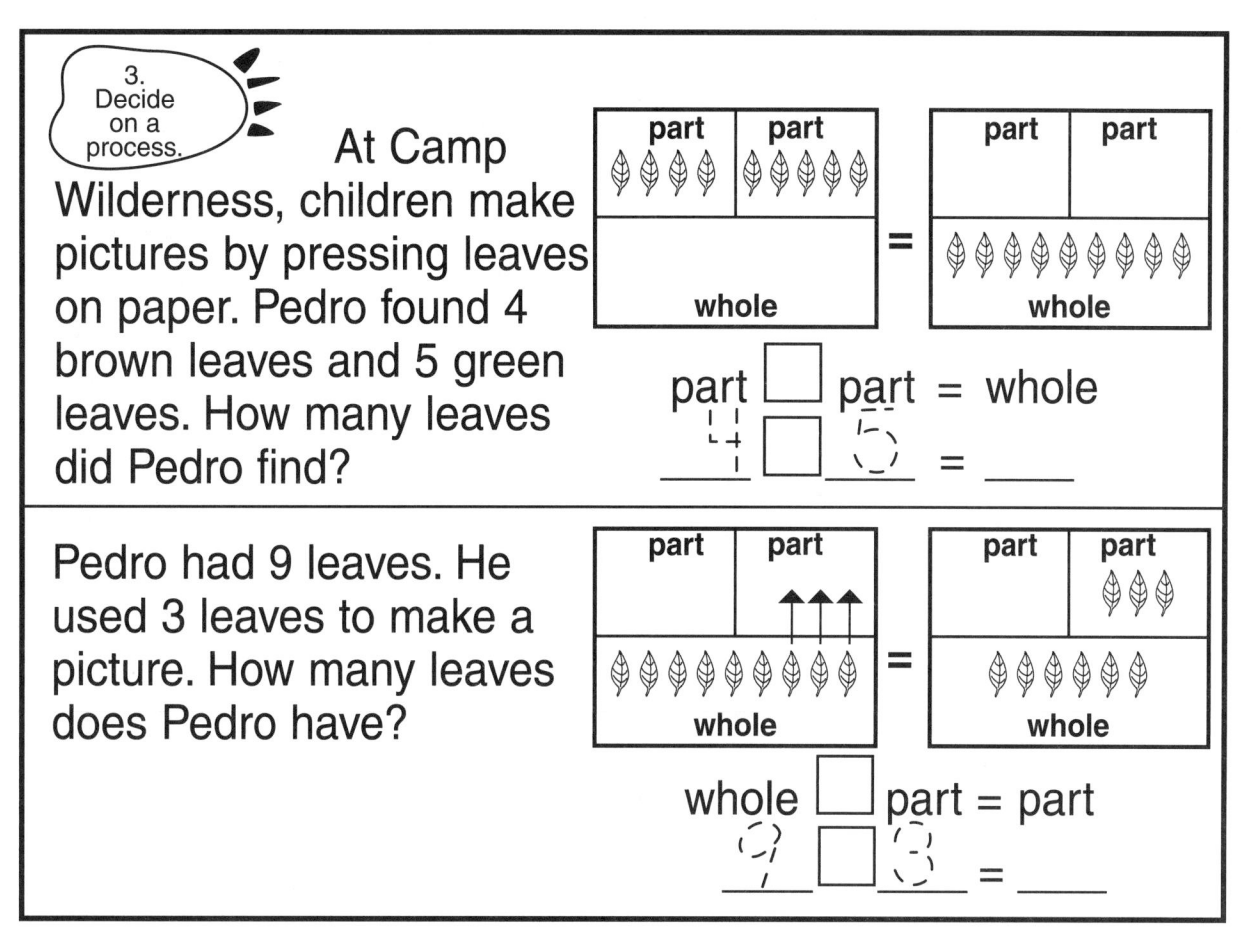

Underline the question. Circle the facts.
Shade the bubble that solves the problem.

1. Meg found 3 yellow leaves and 4 green leaves. How many leaves did Meg find?
   ○ part + part = whole
   ○ whole − part = part

2. Meg had 7 leaves. She gave 5 leaves to Kim. How many leaves does Meg have?
   ○ part + part = whole
   ○ whole − part = part

3. Pat painted 18 pine cones red. She used 6 red pine cones to make a candle holder. How many red pine cones does Pat have?
   ○ 18 + 6   ○ 18 − 6

4. Pat painted 14 pine cones blue. She painted 2 more pine cones blue. How many blue pine cones does Pat have?
   ○ 14 + 2   ○ 14 − 2

Name _____

> Addition and subtraction are opposites.
>
> | Addition puts things together. | Subtraction takes things apart. |
> |---|---|
> | part + part = whole | whole − part = part |
> | 7 + 2 = 9 | 9 − 2 = 7 |
> | The answer is called the sum. | The answer is called the difference. |

Underline the question. Ring the facts. Cross out extra facts. Ring the correct solution.

1. Meg read 8 books. She read 5 more books. How many books did she read in all?
   ○ 8 + 5   ○ 8 − 5

2. Kim read 10 books. 6 of the books were big. How many books were **not** big?
   ○ 10 + 6   ○ 10 − 6

3. Pat had 14 tickets. Pedro had 12 tickets. Pat used 4 tickets. How many tickets does Pat have?
   ○ 14 + 4   ○ 14 − 4
   ○ 12 + 4   ○ 12 − 4

4. Meg read 5 books. 4 of the books were big. She read 2 more books. How many books did she read in all?
   ○ 5 + 4   ○ 5 − 4
   ○ 5 + 2   ○ 5 − 2

5. Find the sum.

     37   61   40
   + 2   + 3   + 3

6. Find the difference.

     47   58   72
   − 3   − 4   − 2

7. Make up a problem about 8 − 2.

_____

_____

Name _____

4. Estimate. 5. Solve. Check back.

Amusement Ride Tickets
"A" ride – 10 tickets
"B" ride – 20 tickets
"C" ride – 30 tickets
All Day Pass – 50 tickets

Kim has 15 tickets.
She wants to take a "B" ride.
Pat gave her 4 more tickets.
Does Kim have enough tickets?

Is 15 + 4 more than 20 or less than 20? My guess is I'll have less than 20 tickets.

```
  15
+  4
  19  tickets
```
Kim has 19 tickets in all.

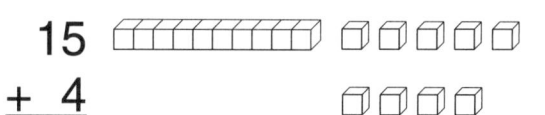

Talk About It:
Was Kim's guess reasonable?
How do you know?
Does Kim have enough tickets for a B ride?   yes   no

Guess. Ring yes or no. Explain.

1. John has 25 tickets.
   He gave 5 tickets to Kim.
   Does John have enough tickets for a "B" ride?

   yes   no

2. Sue has 34 tickets. She got 5 more tickets from Kim.
   Does Sue have enough tickets for an All Day Pass?

   yes   no

3.
```
  32
+  5
```
more than 40   less than 40

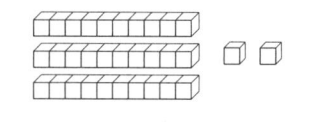

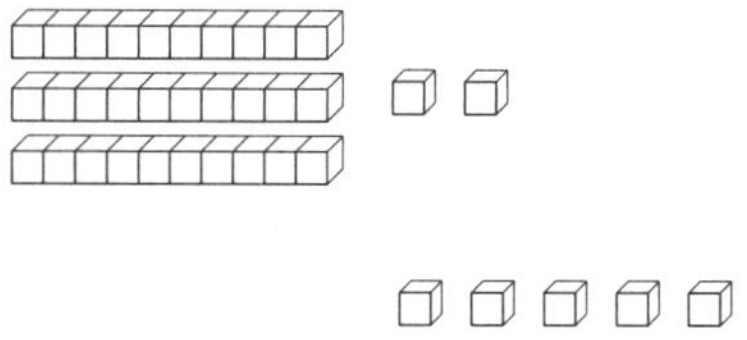

4.
```
  42
+  7
```
more than 50   less than 50

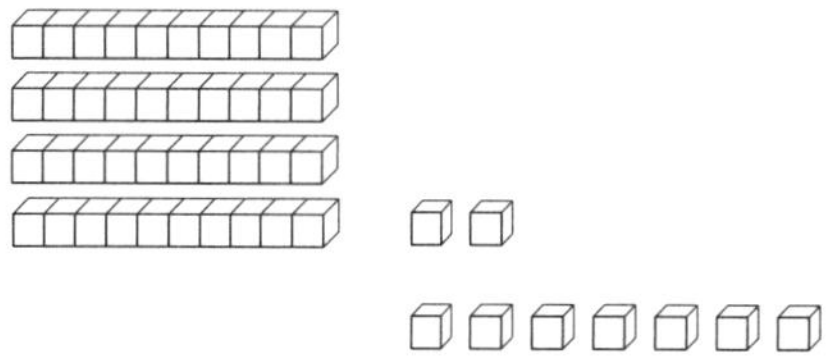

5.
```
    4
+  23
```
more than 30   less than 30

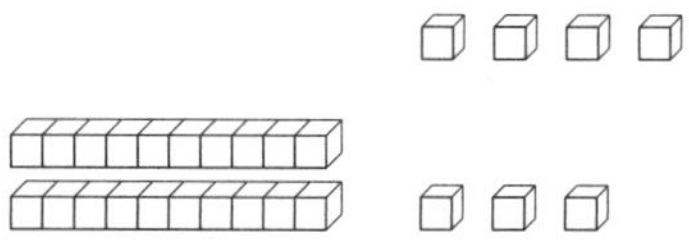

6.
```
  17
+  5
```
more than 20   less than 20

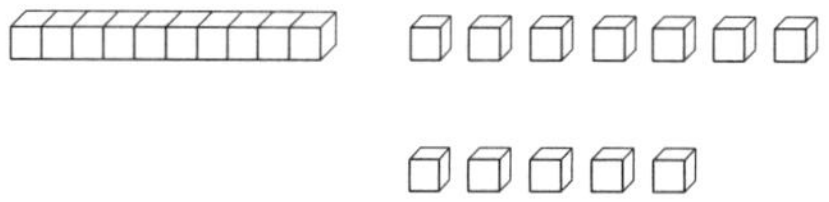

Name _____

Amusement Ride Tickets
"A" ride – 10 tickets
"B" ride – 20 tickets
"C" ride – 30 tickets
All Day Pass – 50 tickets

Pedro has 58 tickets. He wants to buy an All Day Pass and a juice drink. Does he have enough tickets?

Is 58 – 6 more than 50 or less than 50? My guess is more than 50.

```
  58
-  6
  52  tickets
```

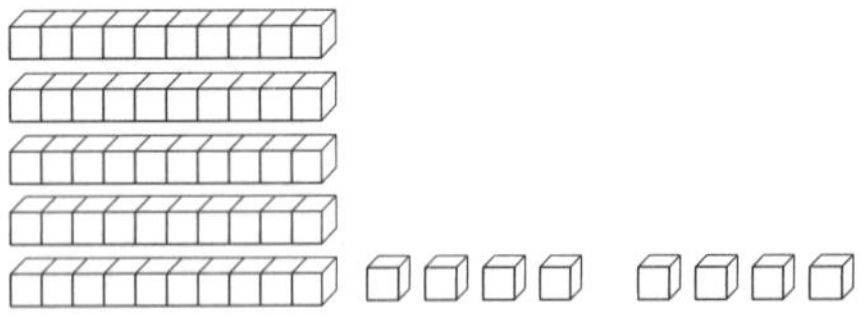

Pedro has 52 tickets left.

Talk About It:

Why did Pedro guess "more than 50"?
Was Pedro's guess reasonable? How do you know?

Guess. Ring yes or no. Explain.

1. Jane has 37 tickets. She bought a juice for 6 tickets. Does she have enough for a "C" ride?

   yes   no

2. Pat has 32 tickets. He bought a juice for 6 tickets. Does he have enough tickets for a "C" ride?

   yes   no

3. 36
   – 4

   more than 30   less than 30

4. 17
   – 6

   more than 10   less than 10

5. 30
   – 2

   more than 20   less than 20

6. 28
   – 5

   more than 20   less than 20

184
© Math Teachers Press, Inc. Reproduction by any means is strictly prohibited.

Name _____

How Many: Fewer, More

Use a model. Draw a picture. Write a number sentence.

1. You have 4 trains and 1 boat. How many fewer boats than trains?

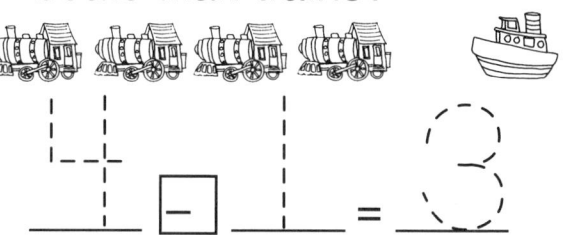

   _4_ □-□ _1_ = _3_

2. You have 7 school buses and 2 fire engines. How many fewer fire engines?

   ____ □-□ ____ = ____

3. You have 9 cars. You have 4 school buses. How many more cars?

   ____ ____ ____

4. You have 6 boats. You have 4 cars. You have 3 trains. How many cars and trains?

   ____ ____ ____

5. You have 9 yellow cubes. You have 7 blue cubes. How many fewer blue cubes?

   ____ ____ ____

6. You have 7 green cubes. You have 3 red cubes. How many more green cubes?

   ____ ____ ____

7. You have 8 △.
   You have 5 □.
   How many fewer □?

   ____ ____ ____

8. You have 3 ○.
   You have 4 □.
   You have 2 △.
   How many □ and △?

   ____ ____ ____

185

Name _____

Favorite Books
Animals 5
Families 3
Fairy tales 4
Humor 2
Trips 1

How many more children chose animal books than fairy tale books?

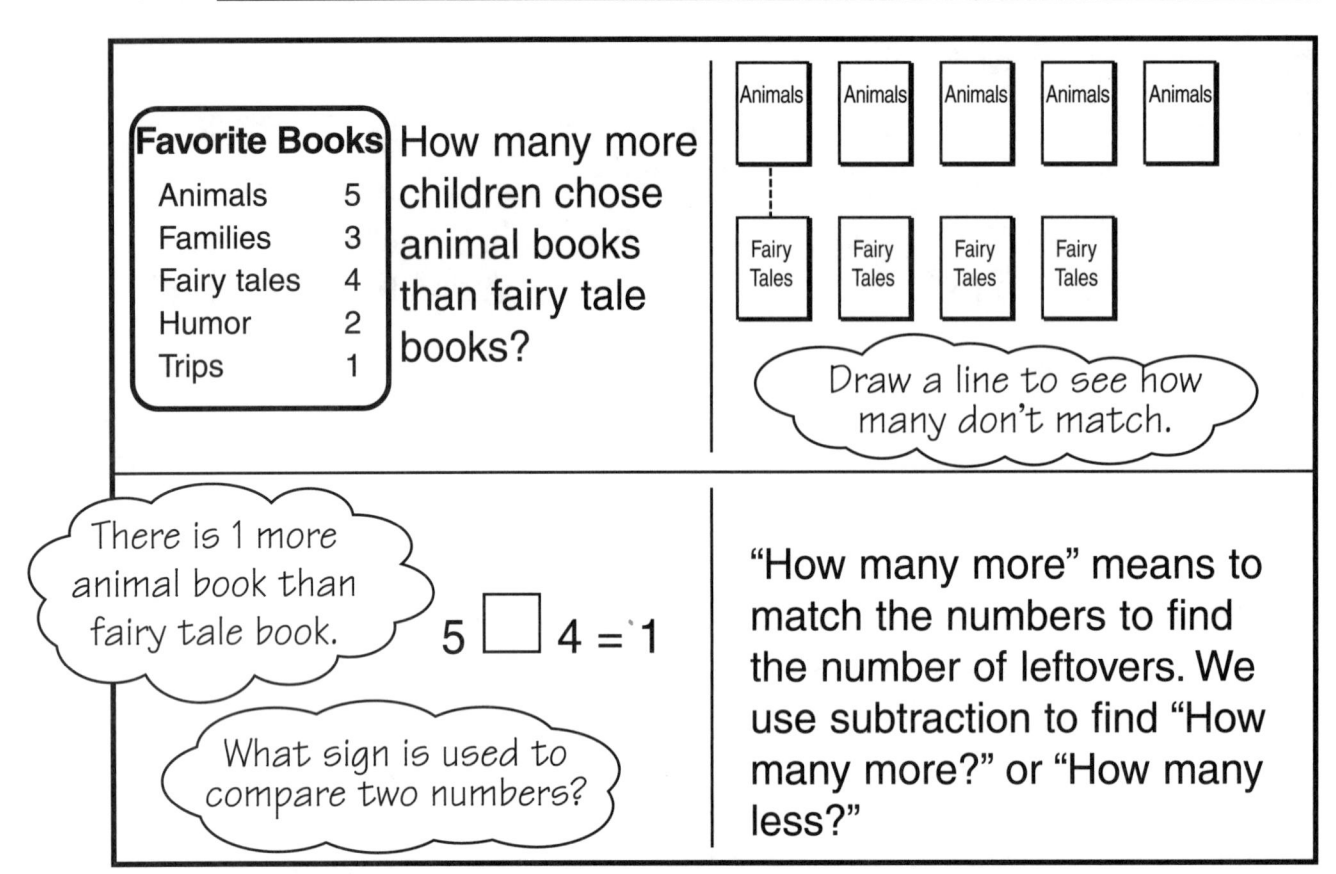

Draw a line to see how many don't match.

There is 1 more animal book than fairy tale book.

5 ☐ 4 = 1

What sign is used to compare two numbers?

"How many more" means to match the numbers to find the number of leftovers. We use subtraction to find "How many more?" or "How many less?"

Draw a picture. Match. Write the number sentence.

1. How many more children chose Fairy Tales than books about humor?

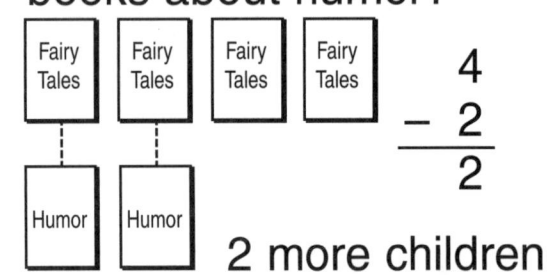

   4
   -2
   $\overline{2}$

   2 more children

2. How many more children chose books about families than books on humor?

   _____

3. How many fewer children chose stories about families than animals?

   _____

4. How many more children chose animal books than humor books?

   _____

Name _____

Adding Three Numbers

You have 2 cubes. Your friend gives you 1 more. You find 4 more. How many cubes do you have?

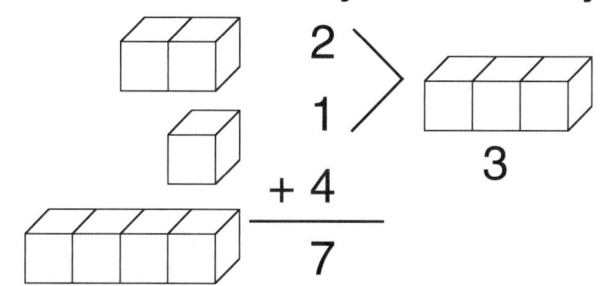

```
  2
  1
+ 4
───
  7
```

Step 1 — Add the first two numbers.

Step 2 — Add the last number.

1.
    5 ⟩ 8   5 ⟩ ──   1 ⟩ ──   6 ⟩ ──   5 ⟩ ──
    3   0   0   4   3
 + 2   + 1   + 3   + 0   + 2
  10

2.
    0 ⟩ ──   1 ⟩ ──   1 ⟩ ──   0 ⟩ ──   2 ⟩ ──
    2   5   3   3   5
 + 4   + 4   + 2   + 6   + 2

3.
    3 ⟩ ──   5 ⟩ ──   2 ⟩ ──   3 ⟩ ──   1 ⟩ ──
    2   1   2   2   7
 + 0   + 4   + 3   + 1   + 2

Name _____

Is There Enough Information?

25 bears are on the train. Some bears got off the train. How many bears are on the train now?

$$\begin{array}{r} 25 \\ -\ ? \\ \hline \end{array}$$

○ enough
○ not enough

The story doesn't tell how many bears got off the train.

---

1. 15 bears are on the train.
   More bears got on the train.
   How many bears are on the train?

   ○ enough
   ○ not enough

2. Pedro has 9 toy trains.
   He has 6 toy cars.
   How many more trains than cars does Pedro have?

   ○ enough
   ○ not enough

3. Meg has 8¢.
   Her mother gave her some more money.
   How much money does Meg have now?

   ○ enough
   ○ not enough

4. Jake had 27 toy boats.
   He gave 5 to Pedro.
   How many boats does Jake have?

   ○ enough
   ○ not enough

5. Mia has 18 marbles.
   Kim has 6 marbles.
   Mia gave Kim some marbles.
   How many marbles does Kim have now?

   ○ enough
   ○ not enough

Name _____

Adding Tens

Pedro wanted to buy tickets for an "A" ride and a "B" ride.
How many tickets should he buy?

BUY TICKETS
"A" ride – 10 tickets
"B" ride – 20 tickets
"C" ride – 30 tickets
All Day Pass – 50 tickets

	Tens	Ones
"A" ride →	▭	
"B" ride →	▭▭	

```
   1 ten
+ 2 tens
_____
____ tens
```

Tens	Ones
1	0
+ 2	0
____	____

Pedro needs 30 tickets.

Use blocks. Find the sum.

1. Meg wants to buy a "B" ride and a "C" ride. How many tickets should she buy? _____ + _____

2. Pat wants to buy an All Day Pass for himself and one for his friend. How many tickets should he buy? _____ + _____

3.   50   20   60   40   50   80
     + 10   + 70   + 10   + 50   + 30   + 10

4.   40   60   20   70   10   60
     + 40   + 30   + 40   + 10   + 40   + 20

5. 50 + 30 = _____   30 + 30 = _____

Name _____

Adding Tens on a Hundred Chart

$\begin{array}{r}10\\+10\\\hline\square\end{array}$ Start at 10. ↓ 1 box.

$\begin{array}{r}16\\+20\\\hline\square\end{array}$ Start at 16. ↓ 2 boxes.

1	2	3	4	5	6	7	8	9	10
11	12	13	14	15	16	17	18	19	20
21	22	23	24	25	26	27	28	29	30
31	32	33	34	35	36	37	38	39	40
41	42	43	44	45	46	47	48	49	50

Use the Hundred Chart to add.

1. $\begin{array}{r}26\\+10\\\hline\end{array}$ Start at _____. ↓ _____ boxes.

2. $\begin{array}{r}14\\+20\\\hline\end{array}$ Start at _____. ↓ _____ boxes.

3. $\begin{array}{r}28\\+20\\\hline\end{array}$ Start at _____. ↓ _____ boxes.

4. $\begin{array}{r}8\\+40\\\hline\end{array}$ Start at _____. ↓ _____ boxes.

5. $\begin{array}{r}27\\+10\\\hline\end{array}$ $\begin{array}{r}34\\+10\\\hline\end{array}$ $\begin{array}{r}15\\+20\\\hline\end{array}$ $\begin{array}{r}18\\+30\\\hline\end{array}$ $\begin{array}{r}25\\+20\\\hline\end{array}$ $\begin{array}{r}30\\+20\\\hline\end{array}$

6. $\begin{array}{r}12\\+30\\\hline\end{array}$ $\begin{array}{r}3\\+40\\\hline\end{array}$ $\begin{array}{r}16\\+30\\\hline\end{array}$ $\begin{array}{r}40\\+8\\\hline\end{array}$ $\begin{array}{r}9\\+30\\\hline\end{array}$ $\begin{array}{r}23\\+20\\\hline\end{array}$

7. 23 + 20 = _____   35 + 40 = _____

8. 18 + 30 = _____   30 + 16 = _____

Name _____

Adding 2 Digit Numbers

24 bears rode the train on Monday. 35 bears rode the train on Tuesday. How many bears rode the train on both days?

TENS	ONES
(blocks showing 2 tens and 3 tens)	(blocks showing 4 and 5)

First, build each number.

TENS	ONES

```
   24
 +35
  ___
  [?]
```

Then, put the ones together, and put the tens together. Read the number.

TENS	ONES

(5) (9) = 59

Now, record what you did with the blocks.

Add the **ones**.
```
  24
+ 35
____
   9
```

Then, add the **tens**.
```
  24
+ 35
____
  59
```

Add.

1.

TENS	ONES		TENS	ONES
1	4		(1 ten)	(4 ones)
+ 3	3		(3 tens)	(3 ones)

( ) ( ) = ____

Name _____

Solve.

1. 53 bears rode the train on Monday. 26 bears rode the train on Tuesday. How many bears rode altogether?

   Guess: _____ Actual: _____

Tens	Ones
5	3
+ 2	6

2. 43 children rode the train on Monday. 35 children rode the train on Tuesday. How many children rode on both days?

   Guess: _____ Actual: _____

Tens	Ones
4	3
+ 3	5

3.
Tens	Ones
3	6
+ 5	2

4.
Tens	Ones
4	0
+ 2	5

5.
Tens	Ones
2	1
+ 7	8

Tens	Ones
3	6
+ 2	1

Tens	Ones
1	4
+ 8	0

6. 43   82   28   13   53   40
   + 55 + 16 + 51 + 50 + 46 + 27

7. Use these steps when you add 2-digit numbers.

   First, add the _____. Then, add the _____.

Name _____

Meg had 50 tickets. She went on a "B" ride. How many tickets does Meg have now?

BUY TICKETS
"A" ride – 10 tickets
"B" ride – 20 tickets
"C" ride – 30 tickets
All Day Pass – 50 tickets

50
− 20

First, build the larger number.

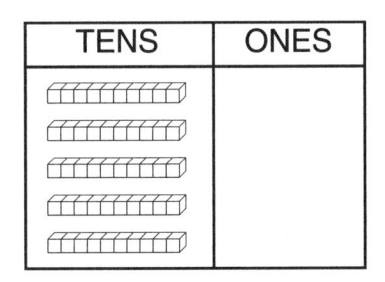

50 | TENS | ONES |

Then, remove the smaller number and read the number left.

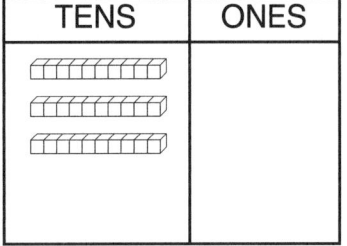

− 20 | TENS | ONES | = 30

30 tickets left

Use blocks. Find the difference.

1. Pat has 80 tickets. He went on a "C" ride. How many tickets does Pat have left?   −_____

2. Pedro has 30 tickets. Pat has 50 tickets. How many more tickets does Pat have?   −_____

3. 40   30   50   60   20   80
   − 10   − 20   − 40   − 20   − 10   − 30

4. 70   90   80   70   90   80
   − 20   − 50   − 40   − 10   − 20   − 20

5. 90 − 30 = _____   80 − 10 = _____

Name _____

Subtracting Tens on a Hundred Chart

```
  20
- 10
 ☐
```
Start at 20.
↑ 1 box.

```
  34
- 20
 ☐
```
Start at 34.
↑ 2 boxes.

1	2	3	4	5	6	7	8	9	10 ↑
11	12	13	14	15	16	17	18	19	20
21	22	23	24	25	26	27	28	29	30
31	32	33	34	35	36	37	38	39	40
41	42	43	44	45	46	47	48	49	50

Use the Hundred Chart to subtract.

1. 26
   − 10
   Start at ____.
   ↑ ____ boxes.

2. 45
   − 20
   Start at ____.
   ↑ ____ boxes.

3. 37
   − 10
   Start at ____.
   ↑ ____ boxes.

4. 42
   − 30
   Start at ____.
   ↑ ____ boxes.

5. 46   38   43   47   39   50
   − 10 − 20 − 30 − 40 − 20 − 30

6. 27   45   75   94   83   80
   − 10 − 40 − 50 − 60 − 70 − 60

7. 38 − 10 = _____   45 − 30 = _____

Name _____

Subtracting 2-digit Numbers

47 bears were on the train. 23 bears got off. How many bears are on the train?

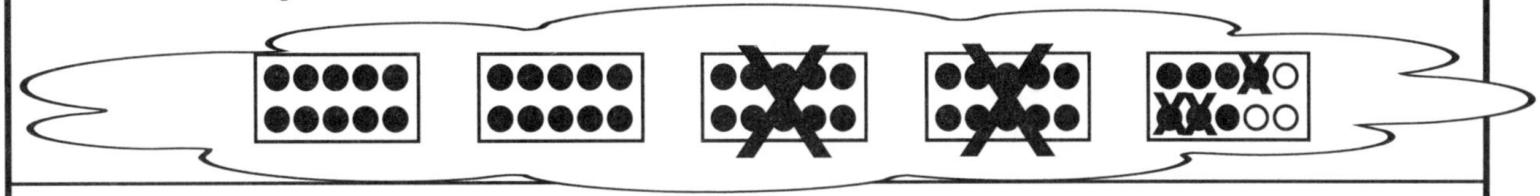

First, build the larger number.

47
− 23

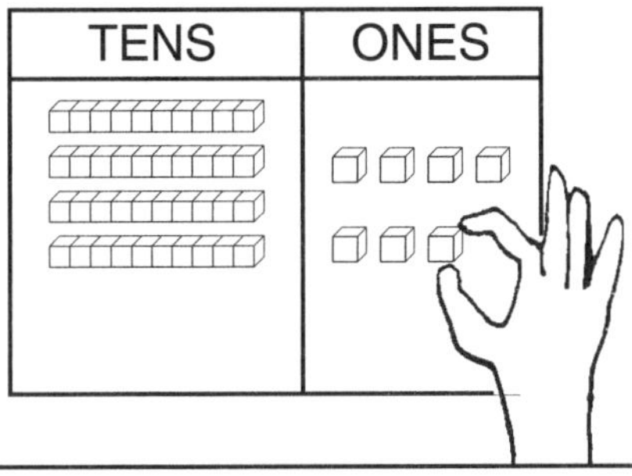

TENS	ONES

Then, remove the smaller number and read the number left.

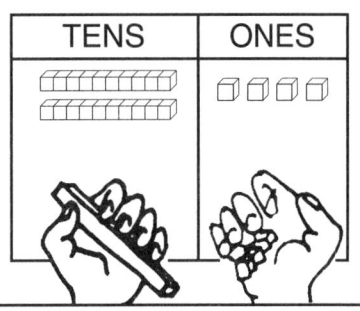

TENS	ONES

24

Now work the problem this way with paper and pencil.

Subtract the **ones**

  4 7
− 2 3
    4

Subtract the **tens**

  4 7
− 2 3
  2 4

Use base ten blocks to subtract. Cross out blocks.

1.   35
  − 12

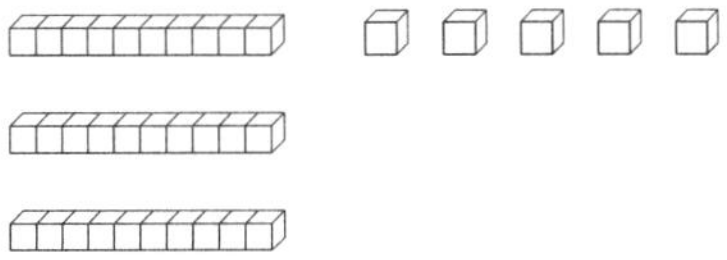

2.   49
  − 35

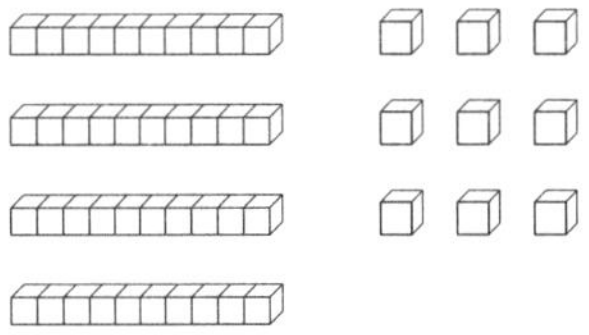

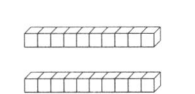

3.   46
  − 20

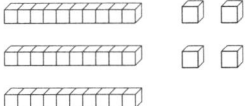

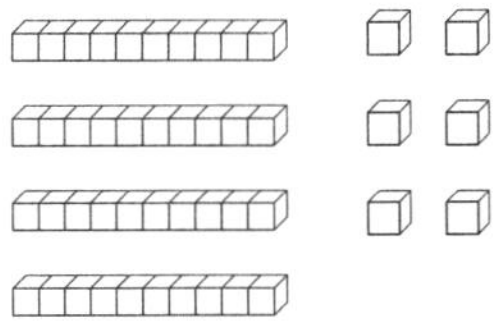

4.   70
  − 30

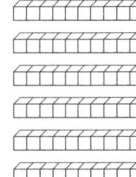

Name _____

Solve.

1. 45 bears are on the plane.
   13 bears get off.
   How many bears are on
   the plane?

Tens	Ones
4	5
−1	3
3	2

   Guess: _____   Actual: _____

2. 56 bears are on the plane.
   24 bears get off.
   How many bears are on
   the plane?

Tens	Ones

   Guess: _____   Actual: _____

Use base ten blocks to subtract. Cross out the blocks subtracted.

3. | Tens | Ones |
   |---|---|
   | 2 | 8 |
   | −1 | 5 |

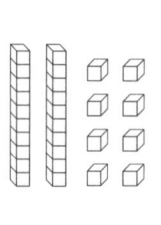

4. | Tens | Ones |
   |---|---|
   | 5 | 8 |
   | −2 | 8 |

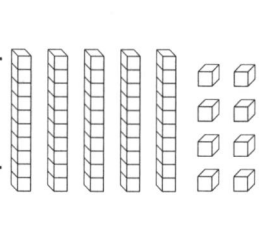

5. 87   76   85   57   75   63
   −35   −56   −42   −40   −32   −21

6. 67   48   65   88   79   56
   −43   −28   −32   −71   −34   −20

7. Use these steps when you subtract 2-digit numbers.
   First, subtract the _____. Then, subtract the _____.

196

Name _____

Adding and Subtracting Money

1. Meg has 35¢. She gets 24¢. How much money does she have?

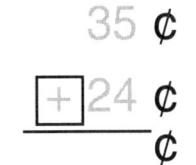

    Guess: _____ ¢

    ```
       35 ¢
    +⃞ 24 ¢
    ───────
        ¢
    ```

2. Pedro has 35¢. He gave 25¢ to Kim. How much does Pedro have?

    Guess: _____ ¢

    ⃞ _____ ¢

3. Jane has 58¢. Joe has 35¢. How much more does Jane have?

    Guess: _____ ¢

    ⃞ _____ ¢

4. Pat earned 75¢. He spent 25¢. How much does Pat have?

    Guess: _____ ¢

    ⃞ _____ ¢

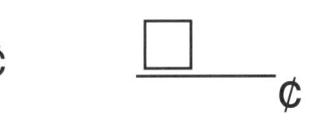

Add or subtract.

5.
```
   33¢      23¢      50¢      46¢      39¢      51¢
+ 26¢    + 25¢    + 40¢    + 42¢    + 30¢    +  8¢
─────    ─────    ─────    ─────    ─────    ─────
```

6.
```
   75¢      29¢      85¢      65¢      74¢      56¢
– 14¢    – 23¢    – 40¢    – 25¢    – 22¢    – 25¢
─────    ─────    ─────    ─────    ─────    ─────
```

7. 22¢ + 26¢ = _____   20¢ + 35¢ = _____

8. 47¢ – 12¢ = _____   66¢ – 20¢ = _____

Name _____

Pat earned 65¢ selling lemonade.
He spent 30¢ and saved the rest.
How much did Pat save?

Which is correct?

```
  65¢            65¢
+ 30¢         −  30¢
─────         ──────
  95¢            35¢
```
(second option circled)

This problem takes money away. So it is subtraction.

Choose the correct answer.

1. Pedro saved 46¢.
   He saved 22¢ more.
   How much did he save?

   ```
     46¢         46¢
   + 22¢       − 22¢
   ─────       ─────
     68¢         24¢
      ○           ○
   ```

2. Kathy had 67¢.
   Her mother gave her 12¢ more.
   How much does she have now?

   ```
     67¢         67¢
   + 12¢       − 12¢
   ─────       ─────
     79¢         55¢
      ○           ○
   ```

3. Brian had 36¢.
   He spent 12¢.
   How much does Brian have?

   ```
     36¢         36¢
   + 12¢       − 12¢
   ─────       ─────
     48¢         24¢
      ○           ○
   ```

4. Deb earned 20¢.
   Pat earned 65¢.
   How much more did Pat earn?

   ```
     65¢         65¢
   + 20¢       − 20¢
   ─────       ─────
     85¢         45¢
      ○           ○
   ```

5. Meg spent 59¢.
   She bought a ball for 20¢.
   What else did she buy?

   30¢   39¢   79¢
    ○   ○   ○

Name _____

Chapter 7 Review

1. How many are there in all?

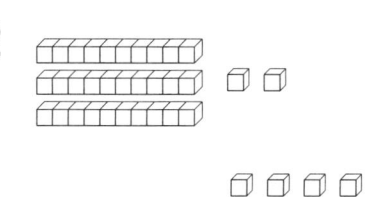

   32
   + 4

2. How many are left? Cross out.

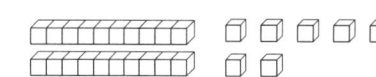

   27
   − 3

3. Find the sum.

   30
   + 50

4. Find the difference.

   80
   − 30

5. Use base ten blocks to add.

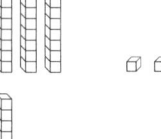

Tens	Ones
4	2
+ 2	3

6. Use base ten blocks to subtract.

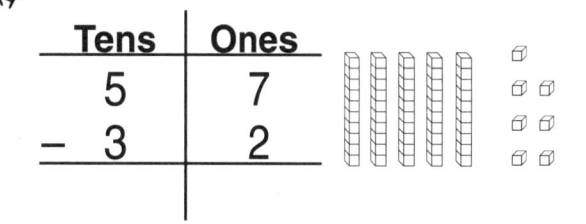

Tens	Ones
5	7
− 3	2

7. Add.

   56
   + 20

   ○ 66
   ○ 67
   ○ 76

8. Subtract.

   75
   − 32

   ○ 33
   ○ 34
   ○ 43

Name _____

9.
```
    5
    1
  + 4
```

10.
```
   12¢
 + 45¢
```

11.
```
   67¢
 - 43¢
```
_____ ¢

12. Ring the best guess.
```
   57
 -  5
```

more than 50   less than 50

13. You have 9 yellow cubes.
You have 7 blue cubes.
How many fewer blue cubes?

____ □ ____ = ____

14. You have 9 cars.
You have 4 school buses.
How many more cars?

____ □ ____ = ____

15. Choose the correct answer.
Mary had 57¢.
Her mother gave her 22¢ more.
How much does she have now?

```
   57¢        57¢
 + 22¢      - 22¢
   79¢        35¢
    ○          ○
```

8
Sums and Differences to 12

Name _____

1. 9 and 🟦 more is ____
2. 8 and 🟦🟦🟦🟦 more is ____
3. 7 and 🟦 more is ____
4. 9 and 🟦🟦🟦 more is ____
5. 8 and 🟦 more is ____
6. 7 and 🟦🟦🟦🟦 more is ____
7. 12 and 🟦 less is ____
8. 12 and 🟦🟦🟦🟦 less is ____
9. 11 and 🟦 less is ____
10. 11 and 🟦🟦🟦🟦 less is ____
11. 11 and 🟦 less is ____
12. 12 and 🟦🟦🟦🟦 less is ____

Name _____

Occupancy: 10

There are 9 🐭.

3 🐭 get on.

How many 🐭?

There will be 1 full car of 10 and 2 more in the next car.

9 + 3 = 12

1. There are 8 🐭.

   4 🐭 get on.

   How many 🐭?

   Draw a picture.

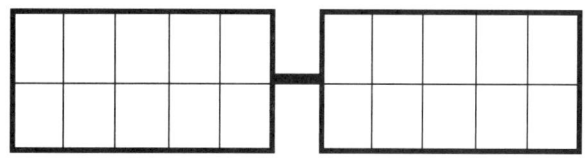

   _____ in all.

2. There are 9 🐭.

   2 🐭 come.

   How many 🐭?

   Draw a picture.

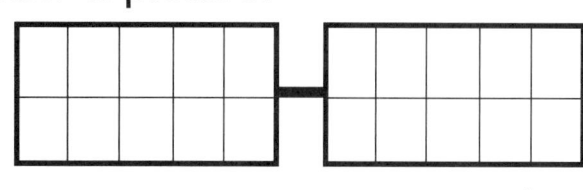

   _____ in all.

3. There are 8 🐭.

   3 🐭 get on.

   How many 🐭?

   Write a number sentence.

   _____ in all.

4. There are 3 🐭.

   9 🐭 come.

   How many 🐭?

   Write a number sentence.

   _____ in all.

Name _____

Mental Math

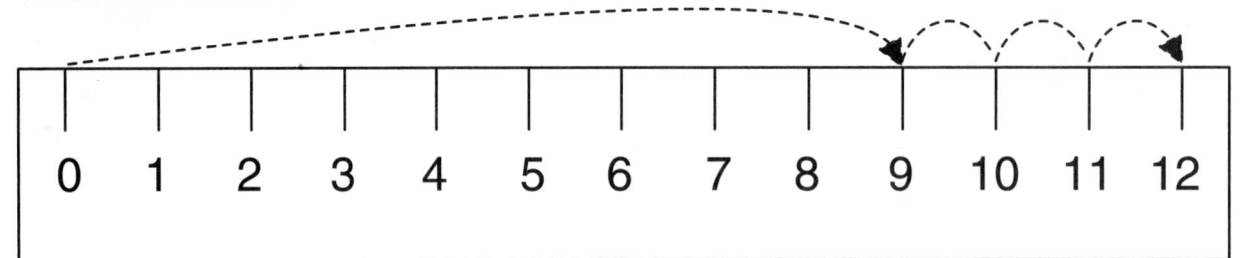

0 1 2 3 4 5 6 7 8 9 10 11 12

3 + 9 =   *Start at the big number. Count up 3.*   9 + 3 = (10, 11, 12) = 12

---

1. 3 + 8 = _____   2 + 9 = _____   3 + 9 = _____

2. 9 + 3 = _____   8 + 3 = _____   9 + 2 = _____

3. If 9 + 1 = 10, then 9 + 2 = _____ and 9 + 3 = _____ .

4. If 8 + 2 = 10, then 8 + 3 = _____ and 8 + 4 = _____ .

5.   8   9   8   9   8   9
     + 2   + 2   + 3   + 3   + 2   + 1

6.   3   4   2   3   1   2
     + 9   + 8   + 9   + 8   + 9   + 8

© Math Teachers Press, Inc., Reproduction by any means is strictly prohibited.

Problem Solving: Write a Number Sentence

7
+ 5
12 in all

7 + 5 = _____

Tell a story. Write a number sentence.

1. There are 8 🐑 .

   4 more 🐑 join them.

   How many 🐑 ?

   8 ____ 4 = _____

2. There are 7 🐴 .

   4 🐴 come.

   How many 🐴 ?

   7 ____ 4 = _____

3. There are 4 🐄 in the barn.

   8 🐄 go in the barn.

   How many 🐄 in the barn?

   _____

4. There are 5 🐖 in the mud.

   7 🐖 join them.

   How many 🐖 in the mud?

   _____

5. You have 5 ¢ .

   You earn 7 ¢ .

   How many ¢ ?

   _____ ¢

6. You have 8 ¢ .

   You find 4 ¢ .

   How many ¢ ?

   _____ ¢

Name _____

Using Models, Mental Math

8 ones + 4 ones   =   12 ones   = 1 ten 2 ones = 12

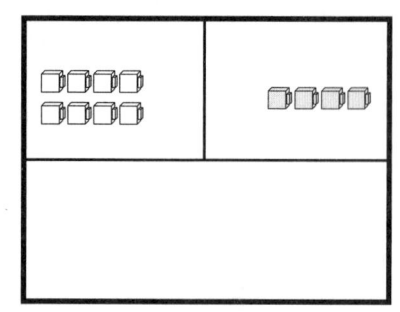

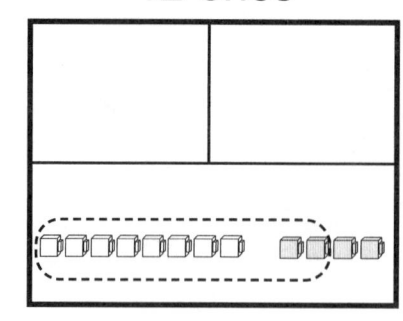

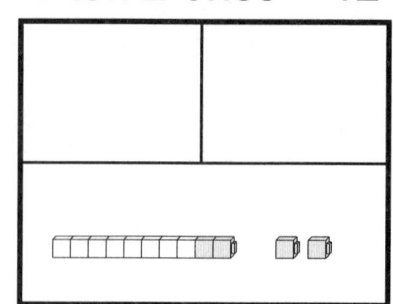

Add using blocks.

1. If 8 + 2 = 10, then 8 + 3 = _____ and 8 + 4 = _____.

2. If 9 + 1 = 10, then 9 + 2 = _____ and 9 + 3 = _____.

3.   7   4   7   8   5   4
     + 4   + 8   + 5   + 4   + 7   + 7

4.   9   8   9   3   3   2
     + 2   + 3   + 3   + 6   + 9   + 9

5. There are 5 🐑.

   7 🐑 come.

   How many 🐑?

6. There are 4 🦆.

   4 🦆 come.

   How many 🦆?

Name _____

Doubles

The double of 6 is 12.

6 + 6 = 12

Ring the doubles. Add.

1.   5   5   5   6   6   7
    + 5   + 6   + 7   + 6   + 5   + 5

2.   4   8   4   7   3   8
    + 7   + 4   + 8   + 4   + 9   + 3

3.   9   6   3   5   5   6
    + 2   + 5   + 8   + 5   + 7   + 6

4. Write number pairs with sums of 11.

   9 , 2   8 , ___   7 , ___   6 , ___

   ___ , ___   ___ , ___   ___ , ___   2 , 9

5. Write number pairs with sums of 12.

   9 , 3   8 , ___   ___ , ___   ___ , ___

   ___ , ___   ___ , ___   ___ , ___   3 , 9

Name _____

Color clue words or signs that tell you to **add.**

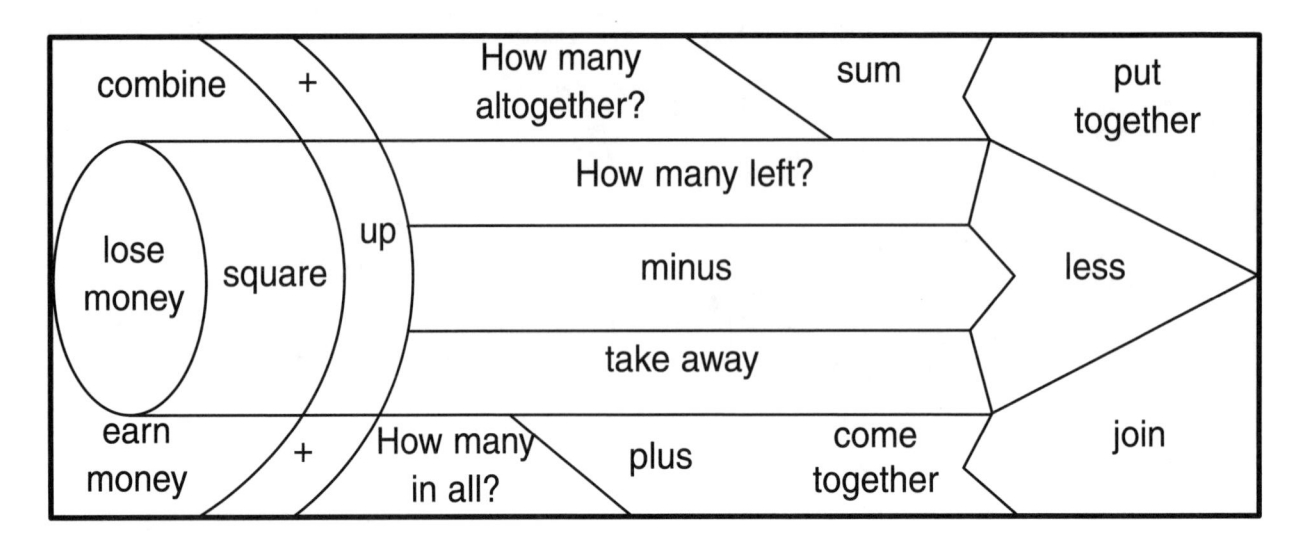

combine + How many altogether? sum put together

lose money square up How many left?

minus

take away less

earn money + How many in all? plus come together join

Sums to 12 No. Correct (15) _____

1.   8   3   8   7   6
   + 4   + 9   + 3   + 4   + 5

2.   6   4   9   5   9
   + 6   + 8   + 2   + 6   + 3

3.   4   3   7   2   5
   + 7   + 8   + 5   + 9   + 7

Name _____

Subtracting 1, 2 or 3

There are 12 🐻.
3 🐻 get off.
How many 🐻?

2 bears get off the second car.
1 bear gets off the first car.

12 − 3 = 9 left

1. There are 11 🐻.
   2 🐻 get off.
   How many 🐻?

   Draw a picture.

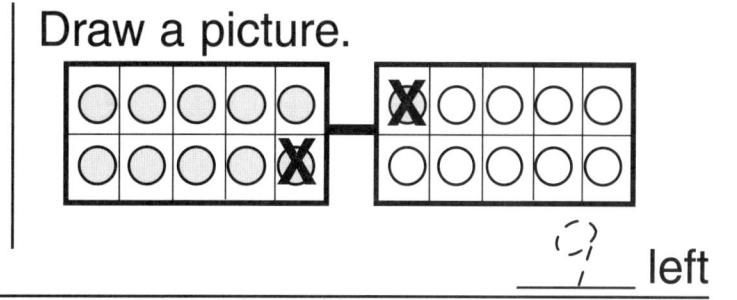

   _9_ left

2. There are 11 🐻.
   8 🐻 get off.
   How many 🐻?

   Draw a picture.

   _____ left

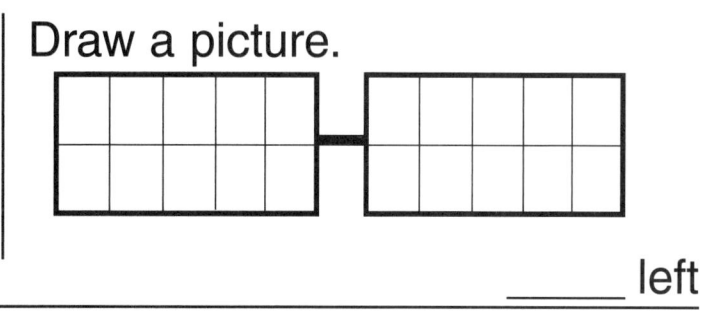

3. There are 11 🐻.
   3 🐻 get off.
   How many 🐻?

   Write a number sentence.

   _____ = _____ left

4. There are 12 🐻.
   3 🐻 get off.
   How many 🐻?

   Write a number sentence.

   _____ = _____ left

Name _____

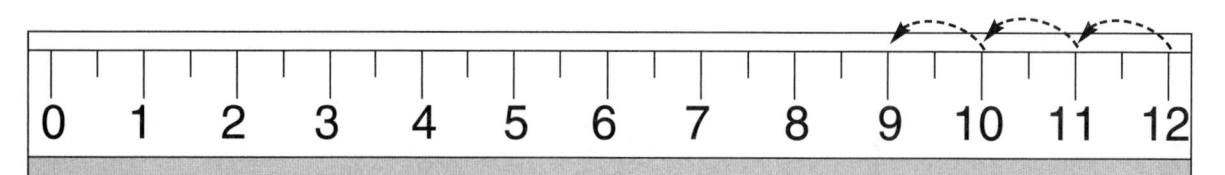

0 1 2 3 4 5 6 7 8 9 10 11 12

12 – 3 = *Start at the big number 12. Count back 3 steps.*   12, 11, 10, 9
12 – 3 = 9

Subtract. Write a new subtraction fact for each fact.

1. If 11 – 2 = 9,   then 11 – 9 = _____

2. If 11 – 3 = 8,   then 11 – 8 = _____

3. 11 – 2 = _____   12 – 3 = _____   11 – 3 = _____

   11 – 9 = 2   _____   _____

4. 12 – 9 = _____   11 – 9 = _____   11 – 8 = _____

   _____   _____   _____

5.   10   11   11   12   10   10
   – 2   – 2   – 3   – 3   – 2   – 1

6.   12   12   11   11   10   10
   – 9   – 8   – 9   – 8   – 9   – 8

210

Name _____

Counting Down in 2 Steps

There are 11 🐻.
4 🐻 get off.
How many 🐻?

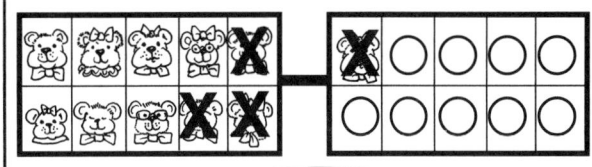

Take 1 bear off the second car to leave 10. Take 3 bears from 10 to leave 7.

11 − 4 = 7 left

1. There are 12 🐻.
   4 🐻 get off.
   How many 🐻?

   Draw a picture.

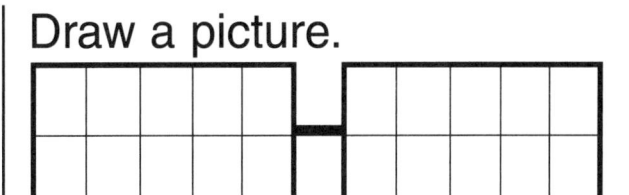

   _____ left

2. There are 12 🐻.
   5 🐻 get off.
   How many 🐻?

   Draw a picture.

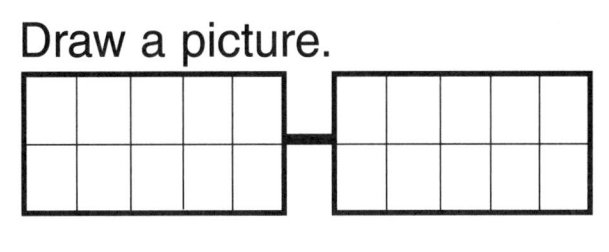

   _____ left

3. There are 11 🐻.
   7 🐻 get off.
   How many 🐻?

   Write a number sentence.

   _____ = _____ left

4. There are 12 🐻.
   8 🐻 leave.
   How many 🐻?

   Write a number sentence.

   _____ = _____ left

Name _____

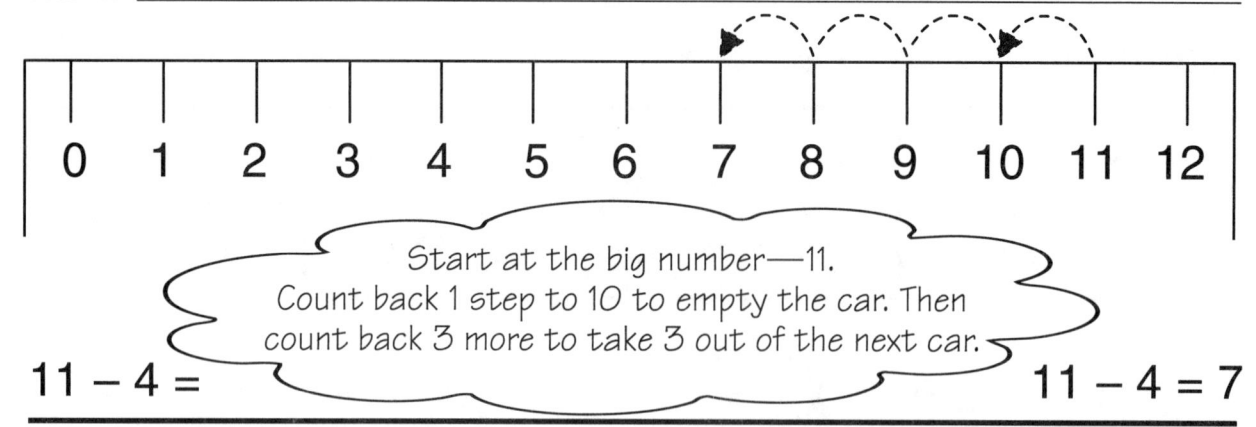

0 1 2 3 4 5 6 7 8 9 10 11 12

11 − 4 =

Start at the big number—11. Count back 1 step to 10 to empty the car. Then count back 3 more to take 3 out of the next car.

11 − 4 = 7

---

1. 11 − 5 = ____   12 − 4 = ____   12 − 5 = ____

    11 − 6 = ____   12 − 8 = ____   12 − 7 = ____

2. 11 − 7 = ____   11 − 3 = ____   12 − 3 = ____

    11 − 4 = ____   11 − 8 = ____   12 − 9 = ____

3. 11   11   12   12   12   12
   − 6   − 4   − 8   − 6   − 7   − 9

4. 11   12   12   11   12   12
   − 5   − 4   − 5   − 7   − 6   − 3

5. Make up a problem about 11 − 6.

   _____

   _____

212

Name _____

Relating Addition and Subtraction

How many in each part of the mat? Subtract to find the missing part.

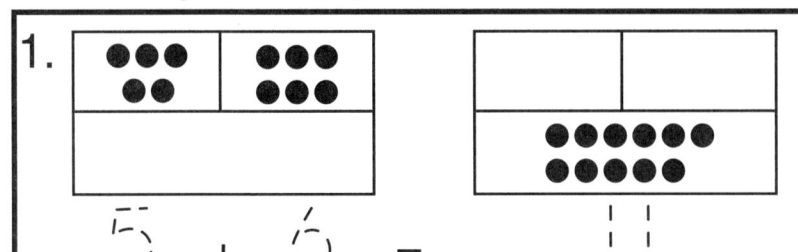

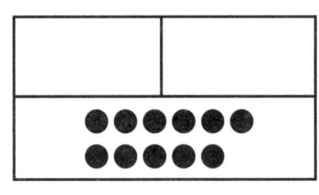

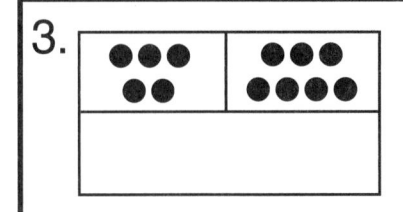

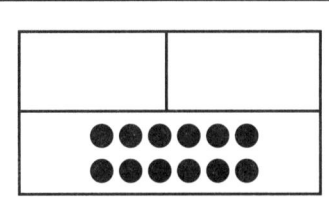

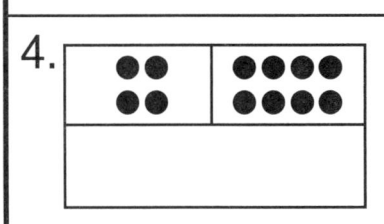

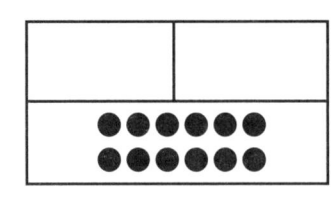

1. ___5___ + ___6___ = _____   11

   11 − 5 = _____

   11 − 6 = _____

2. _____ + _____ = _____

   11 − 9 = _____

   11 − 2 = _____

3. _____ + _____ = _____

   12 − 5 = _____

   12 − 7 = _____

4. _____ + _____ = _____

   12 − 4 = _____

   12 − 8 = _____

5. 11   11   11   11   11   11
   − 6   − 4   − 3   − 7   − 9   − 5

6. 12   12   12   12   12   12
   − 4   − 7   − 6   − 5   − 9   − 8

Name _____

Color clue words or signs that tell you to **subtract**.

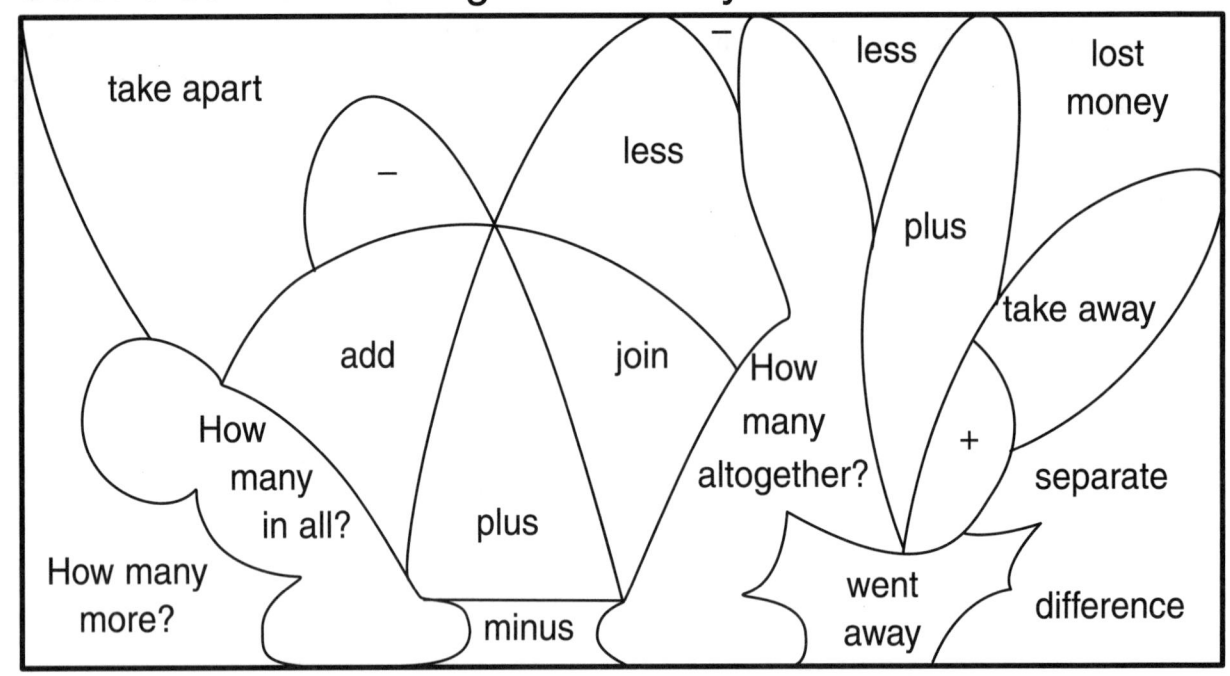

take apart • − • less • less • lost money • add • plus • join • How many altogether? • take away • How many in all? • plus • + • separate • How many more? • minus • went away • difference

Differences to 12　　　　　　　　　　　　No. Correct (15) _____

1.　　11　　　12　　　11　　　12　　　11
　　− 2　　 − 3　　 − 6　　 − 7　　 − 4

2.　　12　　　11　　　12　　　11　　　12
　　− 9　　 − 8　　 − 4　　 − 5　　 − 8

3.　　11　　　12　　　11　　　12　　　11
　　− 7　　 − 5　　 − 4　　 − 6　　 − 3

214
© Math Teachers Press, Inc., Reproduction by any means is strictly prohibited.

Name _____

Names for a Number

What is the fact for the number in the window?

1.

　　　　12

9+2　　8+1　　5+6　　6+6
○　　　○　　　○　　　○

2.

　　　　9

12–4　　12–5　　11–2　　11–3
○　　　○　　　○　　　○

3.

　　　　10

5+6　　6+6　　4+5　　6+4
○　　　○　　　○　　　○

4.

　　　　11

8+4　　9+2　　5+5　　3+9
○　　　○　　　○　　　○

What is the fact for the number in the flower?

5.

　　　　10

5+6　　4+5　　7+4　　7+3
○　　　○　　　○　　　○

6.

　　　　9

5+5　　6+3　　4+4　　5+6
○　　　○　　　○　　　○

7.

　　　　11

7+5　　8+4　　6+6　　8+3
○　　　○　　　○　　　○

8.

　　　　12

8+3　　3+9　　5+5　　4+7
○　　　○　　　○　　　○

Names for a Number

Find other names for each number.
Color the windows with the same name.

12	8+4	7+4	9+3	🐠	6+5	7+5
11	🐈	7+3	5+6	4+8	8+3	6+6
6	11−4	12−6	10−3	11−5	👩	9−2
5	11−6	12−8	10−5	🦜	9−3	12−7
10	5+5	9+0	7+3	2+7	6+4	▭
3	10−6	👦	12−9	3−3	11−8	9−2
4	12−9	11−7	10−6	12−8	9−4	8−4
7	🧑	10−1	12−5	3−3	6−6	9−3

216
© Math Teachers Press, Inc., Reproduction by any means is strictly prohibited.

Name _____

Adding Money

Meg had 9¢.
She earned 3¢.
How much does she have in all?

9¢
+ 3¢
___ ¢

Find the sums.

1. 8¢   3¢   2¢   1¢   8¢   3¢
   + 4¢   + 9¢   + 9¢   + 9¢   + 2¢   + 9¢
   ___ ¢   ___ ¢   ___ ¢   ___ ¢   ___ ¢   ___ ¢

2. 7¢   9¢   6¢   7¢   9¢   4¢
   + 3¢   + 2¢   + 6¢   + 4¢   + 1¢   + 8¢

3. 8¢   5¢   7¢   6¢   2¢   4¢
   + 3¢   + 7¢   + 3¢   + 5¢   + 8¢   + 7¢

4. 4¢   3¢   3¢   6¢   5¢   8¢
   + 5¢   + 7¢   + 8¢   + 3¢   + 2¢   + 2¢

5. Pedro had 7¢.
   He earned 4¢.
   How much does he have in all?
   _____ ¢

6. Pedro wants to buy a whistle for 12¢.
   Does he have enough money?
   Yes   No

Name _____

Subtracting Money

Pedro had 11¢.
He bought a pencil for 5¢.
How much does Pedro have?

5¢

```
   11¢
□ − 5¢
─────
    ¢
```

Find the differences.

1.   11¢   12¢   11¢   10¢   12¢   10¢
     − 2¢   − 4¢   − 4¢   − 7¢   − 8¢   − 6¢
       ¢   ¢   ¢   ¢   ¢   ¢

2.   10¢   11¢   12¢   10¢   11¢   10¢
     − 4¢   − 3¢   − 3¢   − 2¢   − 8¢   − 8¢

3.   12¢   11¢   10¢   12¢   10¢   11¢
     − 9¢   − 5¢   − 9¢   − 5¢   − 1¢   − 6¢

4.   12¢   10¢   11¢   11¢   10¢   12¢
     − 6¢   − 5¢   − 7¢   − 9¢   − 3¢   − 7¢

5. Pedro had 12¢.
   He bought 🖌 for 5¢.
   How much does
   he have left?

   _____ ¢ left

6. Pedro had 12¢.
   Meg had 9¢.
   How much more does
   Pedro have?

   _____ ¢ more

Name _____

Chapter 8 Review

1. Count on to add.

   $\boxed{9} + 2$

   10   11   12
   ○   ○   ○

2. Count on to add.

   $\boxed{8} + 3$

   10   11   12
   ○   ○   ○

3. Count back to subtract.

   $\text{⑫} - 3$

   7   8   9
   ○   ○   ○

4. Count back to subtract.

   $\text{⑪} - 3$

   8   9   10
   ○   ○   ○

5. Add.

   $\begin{array}{r} 6 \\ + 6 \\ \hline \end{array}$

6. Add.

   $\begin{array}{r} 7 \\ + 4 \\ \hline \end{array}$

7. Subtract.

   $\begin{array}{r} 12 \\ - 7 \\ \hline \end{array}$

8. Subtract.

   $\begin{array}{r} 11 \\ - 9 \\ \hline \end{array}$

Name _____

9. There are 8 🐹.
   3 🐹 get on.
   How many 🐹?

Write a number sentence.

_____ ☐ _____ = _____

---

10. There are 12 🐹.
    3 🐹 get off.
    How many 🐹?

Write a number sentence.

_____ = _____ left

---

11. Add.

    8¢
  + 4¢
  _____
    ¢

12. Subtract.

    11¢
  − 5¢
  _____
    ¢

What is the fact for the number in the flower?

13.

(11)

5+5   6+3   4+4   5+6
 ○   ○   ○   ○

14.

(8)

12−3   11−2   11−3   10−3
 ○   ○   ○   ○

Choose the correct number sentence.

15. You have 8 🪙.
    You find 4 🪙.
    How many 🪙?

    ○ 8¢ + 4¢ = 12¢
    ○ 8¢ − 4¢ = 4¢

220
© Math Teachers Press, Inc., Reproduction by any means is strictly prohibited.

9
Fractions and Chance

Name _____

Show fair shares.

1.	6		
2.	8		

3.	6			
4.	12			
5.	3			

6.	8				
7.	4				
8.	12				

Name _____

Share the Snacks
Draw lines or circles to help.

1. 2 children. 4 🐟.

   Each gets _____ 🐟.
   _____ left over.

2. 2 children. 5 🍪.

   Each gets _____ 🍪.
   _____ left over.

🍎			left over
3. 6			
4. 5			
5. 7			

🍌					left over
6. 8					
7. 6					
8. 12					

Name _____

Equal Parts

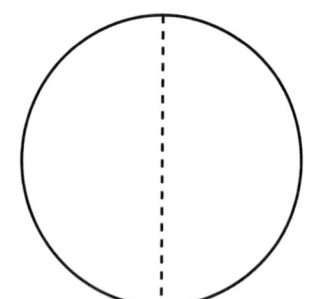

This circle has ____ equal parts.

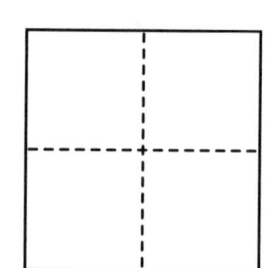

This square has ____ equal parts.

Ring the shapes that have equal parts.

1.

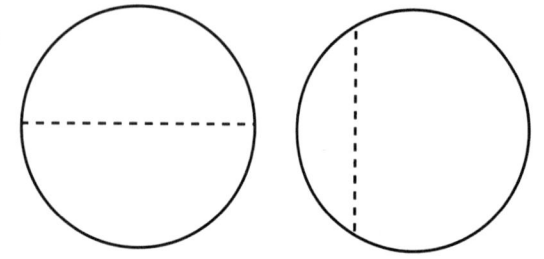

2.

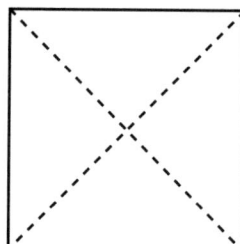

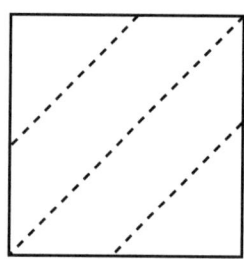

3.

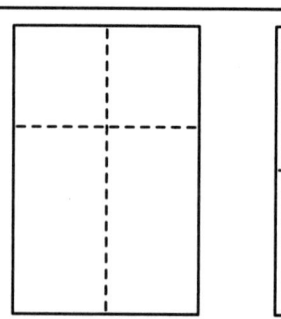

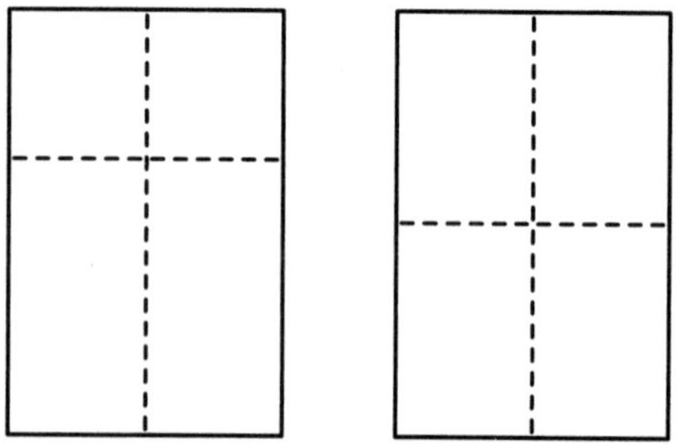

4.

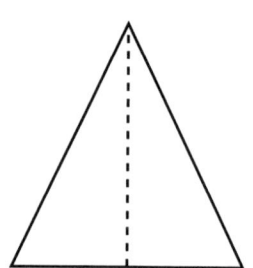

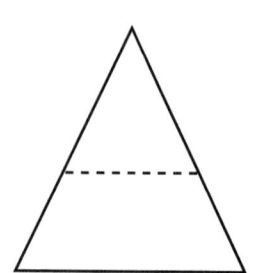

Write the number of equal parts.

5. ____

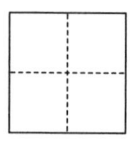

6. ____

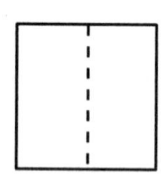

7. ____

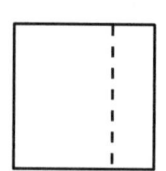

8. ____

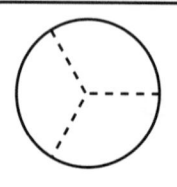

9. ____

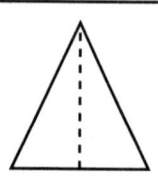

10. ____

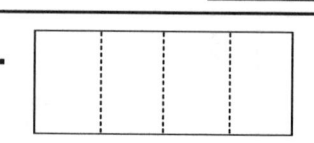

224
© Math Teachers Press, Inc. Reproduction by any means is strictly prohibited.

Name _____

One Half

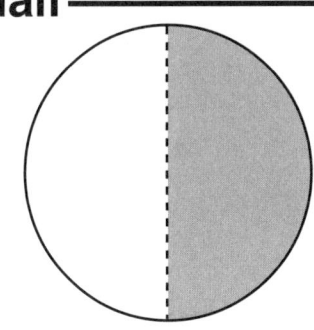

There are 2 equal parts.

1 of 2 parts is shaded.

$\frac{1}{2}$ (one half) is shaded.

Ring the shapes that show one half.

1. 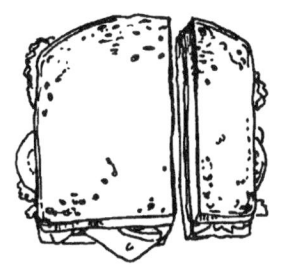   2. 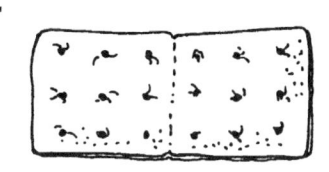   3.

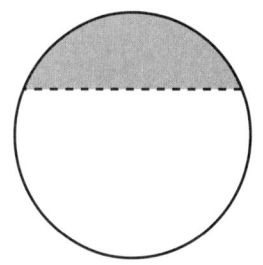

4. 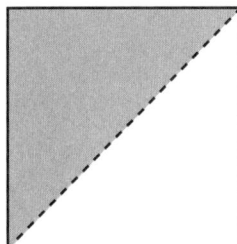   5. 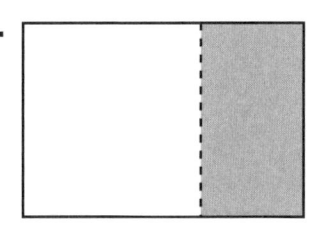   6.

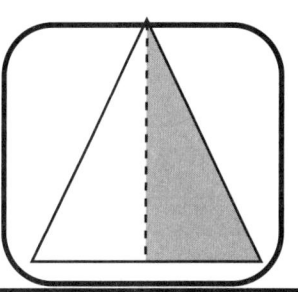

Color $\frac{1}{2}$.

7. 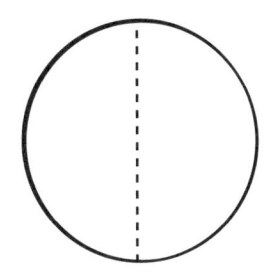   8. 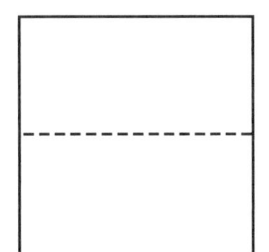   9.

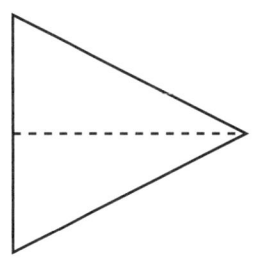

10. 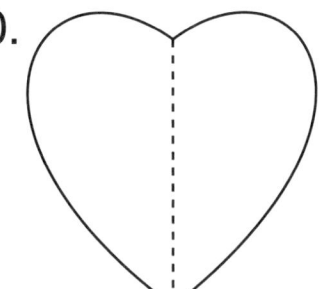   11. 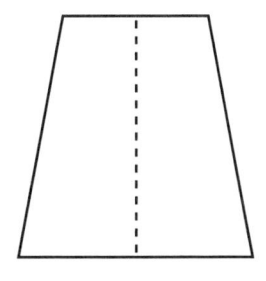   12.

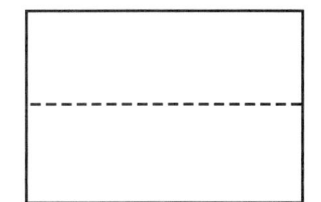

Name _____

Write a fraction.

1.

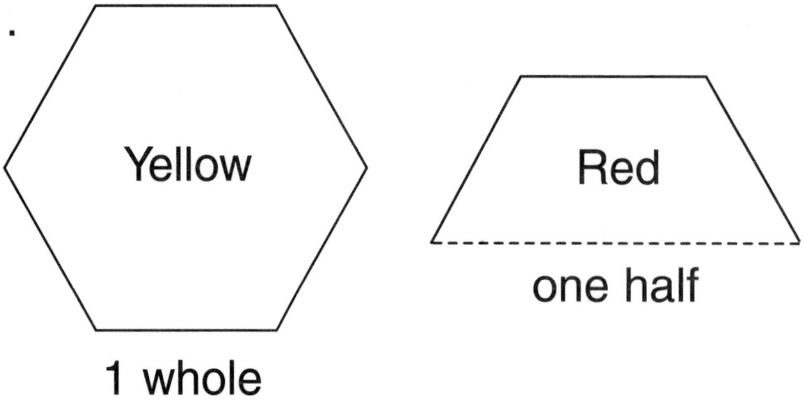

Yellow
1 whole

Red
one half

Red = $\dfrac{1}{2}$

Draw a line on each figure to show matching parts.

2.   3.    4.

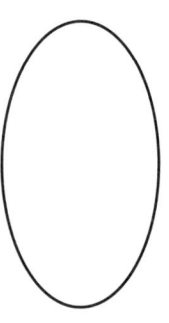

One half is shown. Draw the whole shape.

5.    6. 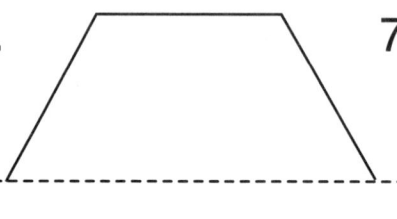   7.

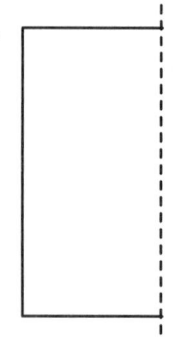

Use string to measure the line. Fold.
Mark one half on the line.

8.

226

Name _____

One Fourth

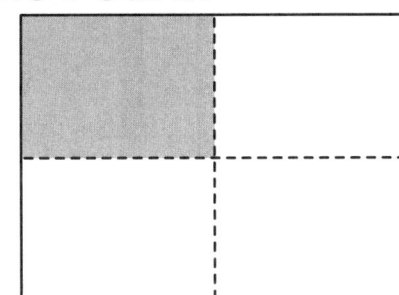

There are 4 equal parts.

1 of 4 parts is shaded.

$\frac{1}{4}$ (one fourth) is shaded.

Ring the shapes that show one fourth shaded.

1. 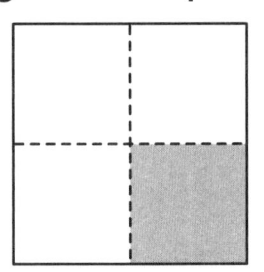
2. 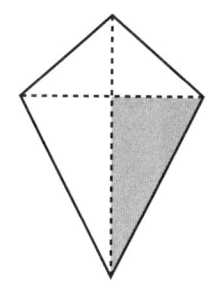
3. 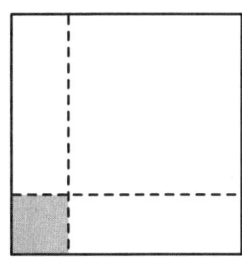
4. 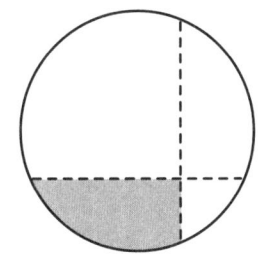
5. 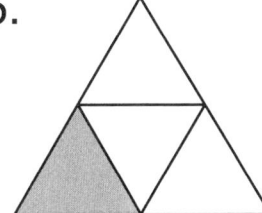
6.

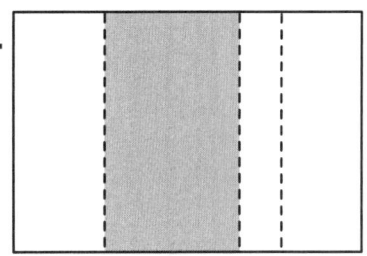

Color $\frac{1}{4}$.

7. 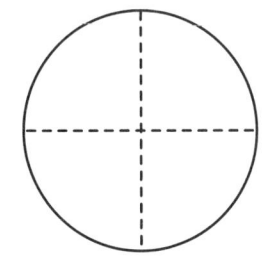
8. 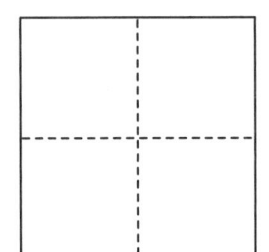
9. 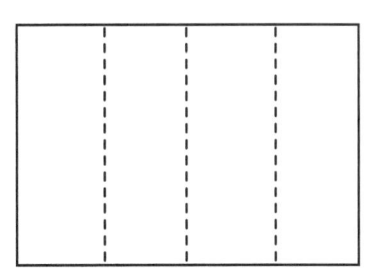
10. 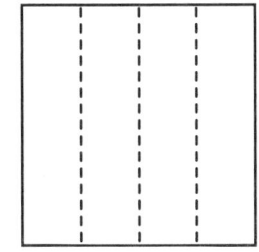
11. 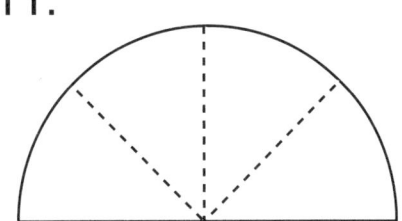
12.

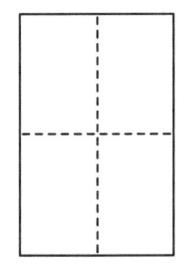

227
© Math Teachers Press, Inc. Reproduction by any means is strictly prohibited.

Name _____

Draw lines on each figure to show 4 equal parts.

1. 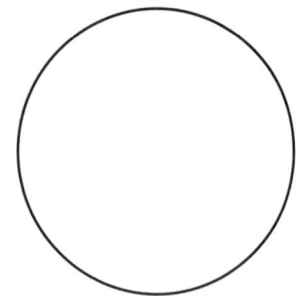   2.   3.

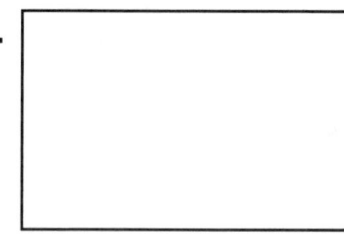

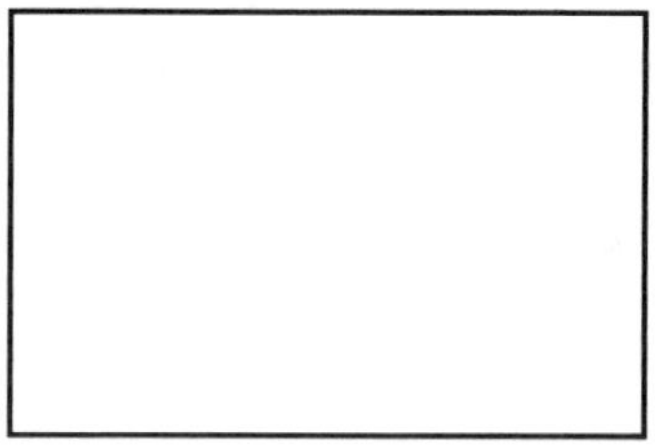

Color $\frac{1}{4}$ blue. Color $\frac{1}{4}$ yellow. Color $\frac{1}{4}$ red. Color $\frac{1}{4}$ green.

4. 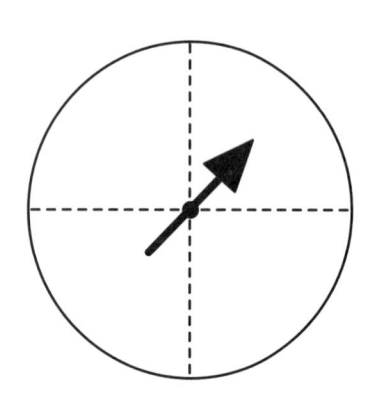   5.

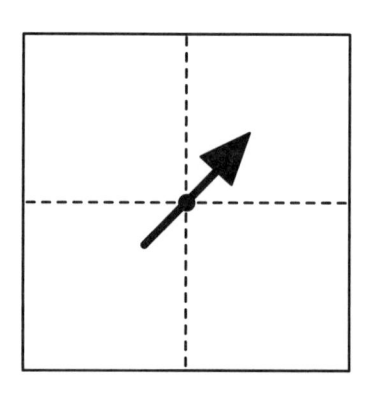

6.    7.

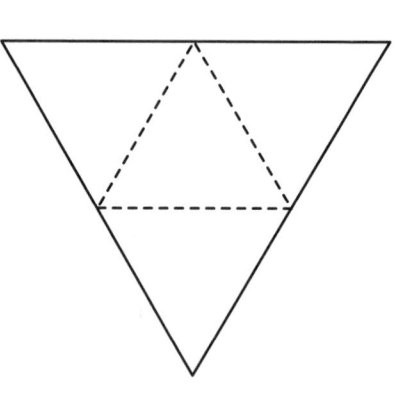

8. Cut. Draw dotted lines to show how to fold this strip into one fourths.

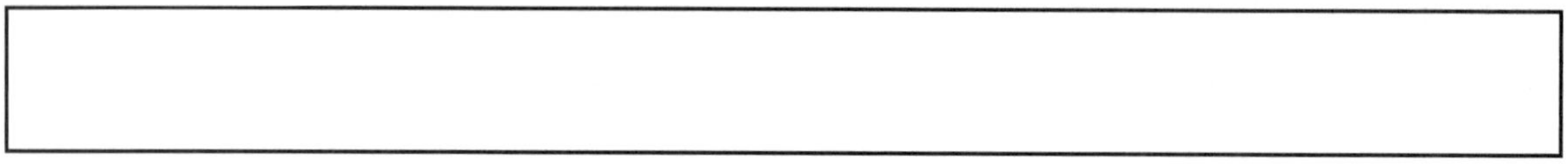

Name _____

One Third

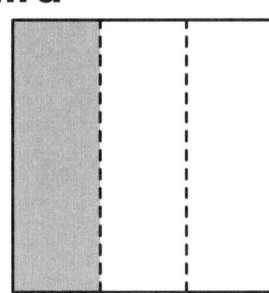

There are 3 equal parts.

1 of 3 parts is shaded.

$\frac{1}{3}$ (one third) is shaded.

Ring the shapes that show one third.

1. 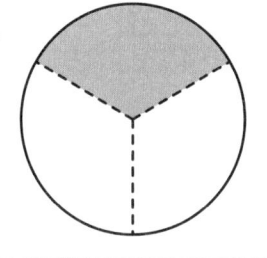
2. 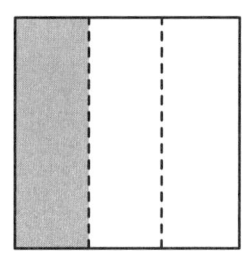
3. 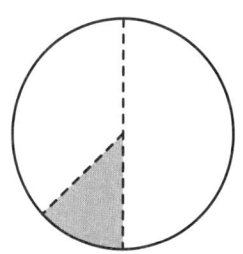
4. 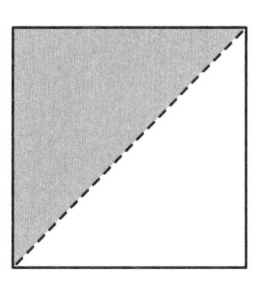
5. 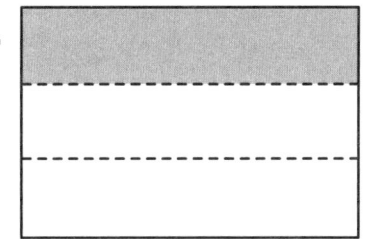
6.

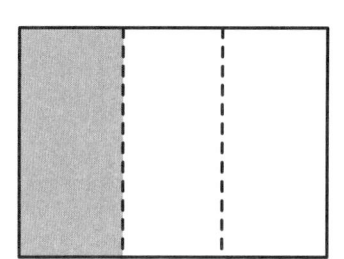

Color $\frac{1}{3}$.

7. 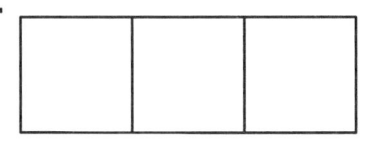
8. 
9. 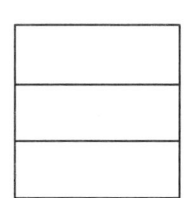
10. 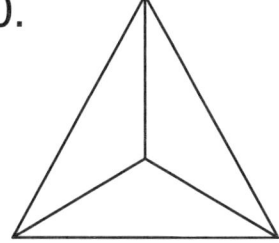
11. 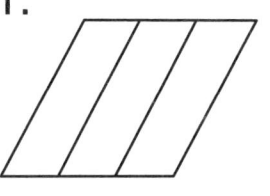
12.

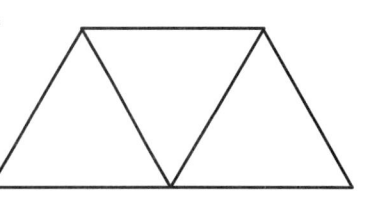

229
© Math Teachers Press, Inc. Reproduction by any means is strictly prohibited.

Name _____

Ring $\frac{1}{2}$, $\frac{1}{3}$, or $\frac{1}{4}$.

1. $\frac{1}{2}$ $\frac{1}{3}$ $\frac{1}{4}$	2. $\frac{1}{2}$ $\frac{1}{3}$ $\frac{1}{4}$	3. $\frac{1}{2}$ $\frac{1}{3}$ $\frac{1}{4}$
4. $\frac{1}{2}$ $\frac{1}{3}$ $\frac{1}{4}$	5. $\frac{1}{2}$ $\frac{1}{3}$ $\frac{1}{4}$	6. $\frac{1}{2}$ $\frac{1}{3}$ $\frac{1}{4}$
7. $\frac{1}{2}$ $\frac{1}{3}$ $\frac{1}{4}$	8. $\frac{1}{2}$ $\frac{1}{3}$ $\frac{1}{4}$	9. $\frac{1}{2}$ $\frac{1}{3}$ $\frac{1}{4}$

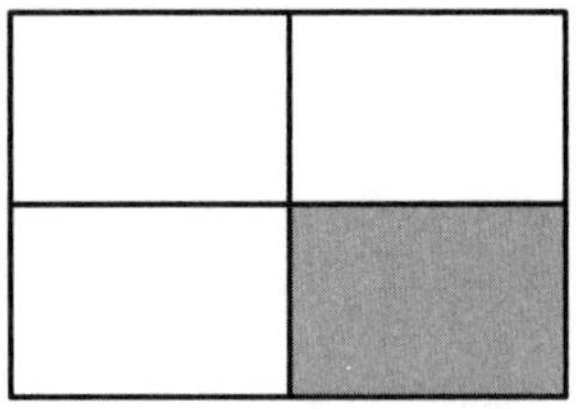

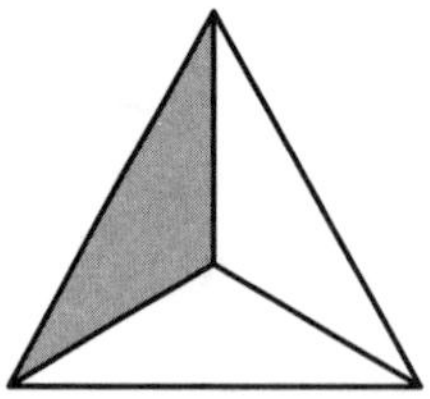

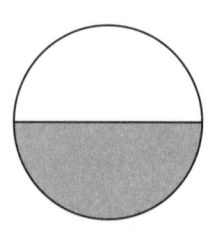

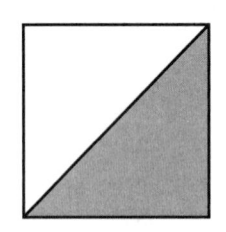

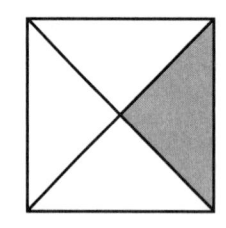

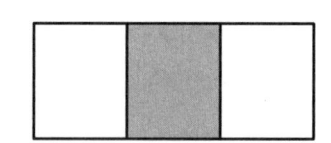

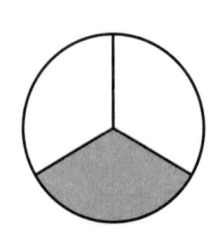

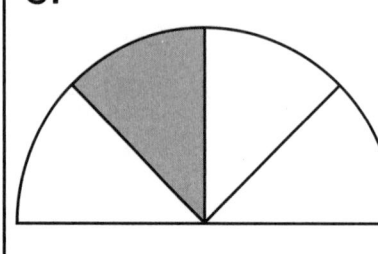

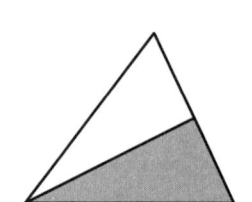

Write the fraction for the shaded part.

10. _____   11. _____   12. _____

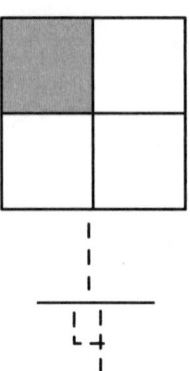

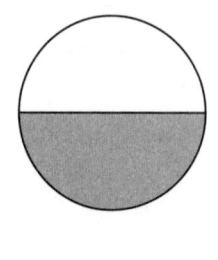

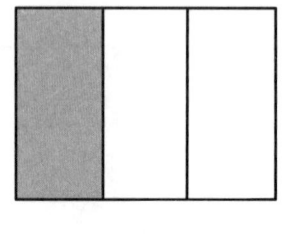

Name _____

Fractions as Part of a Set

Build the train. Color the cubes. What fraction is red? Ring the fraction.

1. [R|Y|B|G]

   $\frac{1}{2}$ $\frac{1}{3}$ $\frac{1}{4}$

2. [R] [Y] [B] [G]

   $\frac{1}{2}$ $\frac{1}{3}$ $\frac{1}{4}$

3. [R|Y]

   $\frac{1}{2}$ $\frac{1}{3}$ $\frac{1}{4}$

4. [R] [Y]

   $\frac{1}{2}$ $\frac{1}{3}$ $\frac{1}{4}$

5. [R|Y|B]

   $\frac{1}{2}$ $\frac{1}{3}$ $\frac{1}{4}$

6. [R] [Y] [B]

   $\frac{1}{2}$ $\frac{1}{3}$ $\frac{1}{4}$

7. Color $\frac{1}{2}$.

8. Color $\frac{1}{3}$.

9. Color $\frac{1}{4}$.

10. Think about it:

    $\frac{1}{2}$ of the bears are in the lake.

    $\frac{1}{2}$ of the bears are in the tent.

    How many bears in all? _____

Name _____

Yes, No, Maybe

A bag is filled with triangles. You pick 1 shape from the bag. Will you pick a triangle? Why?

yes   maybe   no
○   ○   ○

Yes—certain to happen

A bag is filled with triangles. You pick 1 shape from the bag. Will you pick a square? Why?

yes   maybe   no
○   ○   ○

No—impossible

A bag is filled with triangles and squares. You pick a shape.
Will it be a triangle?
yes   maybe   no
○   ○   ○

Will it be a square?
yes   maybe   no
○   ○   ○

Maybe you will pick a square. Maybe you will pick a triangle. An uncertain event is a chance event.

Maybe—uncertain, chance

Will it happen?

1. You throw a ball up in the air. Will it come down?

   yes   maybe   no
   ○   ○   ○

2. You try to jump to the moon. Will you do it?

   yes   maybe   no
   ○   ○   ○

3. You pick a shape. Will it be a circle?

   yes   maybe   no
   ○   ○   ○

4. You pick a shape. Will it be a circle?

   yes   maybe   no
   ○   ○   ○

Name _____

Are the Chances Fair?

1. You and your friend play a game with a spinner. You move 1 space when the spinner lands on green. Your friend moves 1 space when the spinner lands on yellow. The first person to move 25 spaces is the winner.

    Spin and tally until one player has 25 points.

		Total No.
green		
yellow		

Is the game fair?   yes   maybe   no
                     ○   ○   ○

2. You play another game with this spinner.
    Spin and tally until one player gets 25 points.

    Is the game fair?
    yes   maybe   no
    ○   ○   ○

		Total No.
green		
yellow		

3. Which game is fair? Why?

_____

_____

233

Name _____

Which Game is Fair?

1. Pick a bear. Put the bear back.
   The first player to get 10 points wins.

	Game 1	Game 2	Game 3	Game 4
Green				
Yellow				

   Green won ____ times. Yellow won ____ times.
   Do you think the game is fair?   Yes   No

2. Pick a bear. Put the bear back.
   The first player to get 10 points wins.

	Game 1	Game 2	Game 3	Game 4
Green				
Yellow				

   Green won ____ times. Yellow won ____ times.
   Do you think the game is fair?   Yes   No

3. Draw and color pictures of bears in a bag for a fair game.

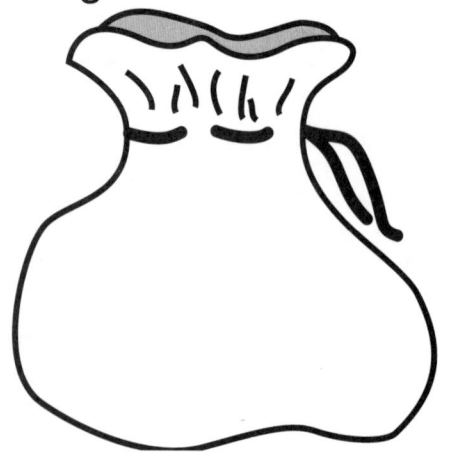

4. Draw and color pictures of bears in a bag for an unfair game.

   _____ will win most.

234
© Math Teachers Press, Inc. Reproduction by any means is strictly prohibited.

Name _____

Are the spinners fair?

1. Spin. Tally the number the spinner lands on. Which number gets 5 points first?

1	2	3	4	5	6

Do you think the game is fair?   Yes   No

2. Spin. Tally the number the spinner lands on. Which number gets 5 points first?

1	2	3	4	5

Do you think the game is fair?

Why? _____

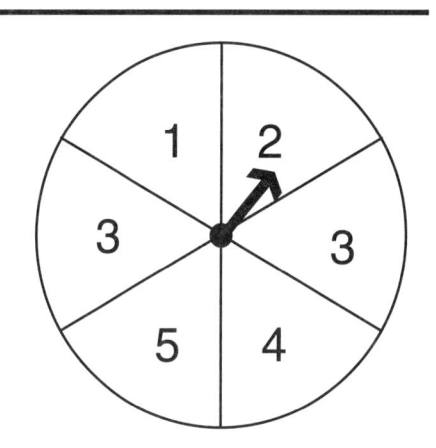

3. Throw 2 dice. Find the sum. Tally. Which sum is the first to come up 5 times? Ring the sum.

Sums of 2 dice: Game A										
2	3	4	5	6	7	8	9	10	11	12

Sums of 2 dice: Game B										
2	3	4	5	6	7	8	9	10	11	12

Name _____

How Many Different Ways?

1. You have a bag with 1 blue cube, 1 yellow cube and 1 red cube inside. You select 1 cube at a time. How many different ways could you pick a cube?

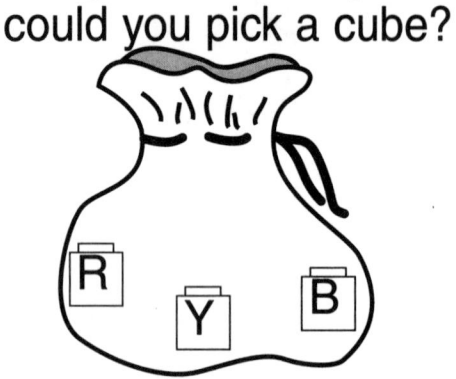

   Color the different ways:

	1st color	2nd color	3rd color
1.	R	Y	B
2.			
3.			
4.			
5.			
6.			

2. Help dress the bears. The bear has a red shirt and a blue shirt. The bear has yellow pants and green pants. How many different outfits can the bear wear? Color the different outfits.

   _____ outfits

Name _____

Chapter 9 Review

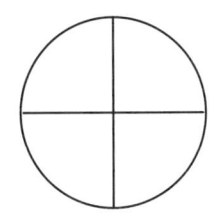

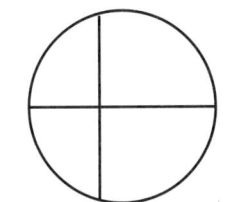

1. Ring the figure with equal parts.

2. How many equal parts are there?

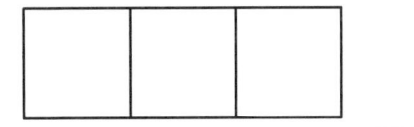

    _____

3. Ring the picture that shows halves.

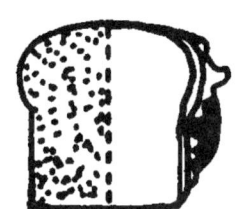

4. Ring the picture that shows the fraction $\frac{1}{4}$.

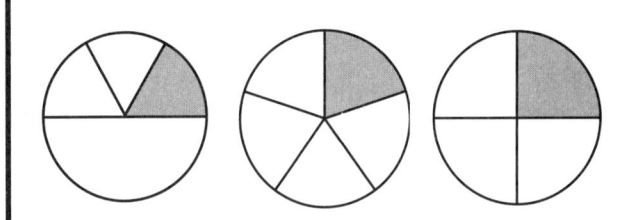

5. Shade $\frac{1}{3}$.

6. Shade $\frac{1}{2}$.

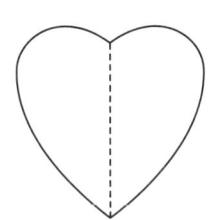

7. Ring the fraction that names the shaded part.

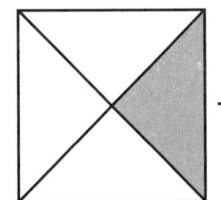

    $\frac{1}{2}$ $\frac{1}{3}$ $\frac{1}{4}$

8. Ring the fraction that names the shaded part.

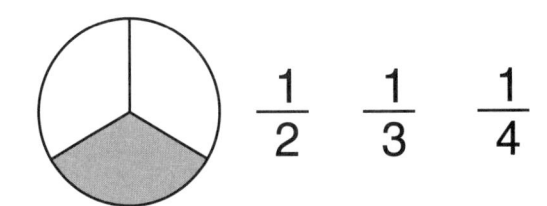

    $\frac{1}{2}$ $\frac{1}{3}$ $\frac{1}{4}$

Name _____

9. Ring the fraction that names the shaded part of the set.

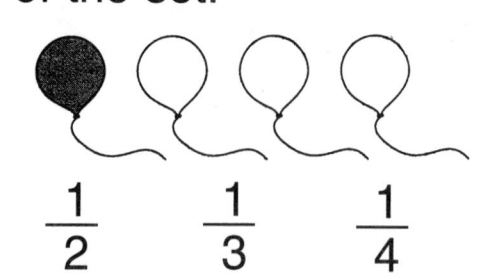

$\frac{1}{2}$ $\frac{1}{3}$ $\frac{1}{4}$

10. Color $\frac{1}{3}$.

11. You pick a shape. Will it be a circle?

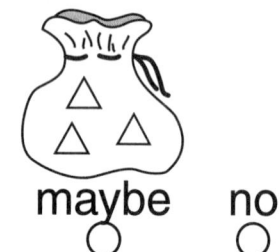

yes   maybe   no

12. You pick a shape. Will it be a circle?

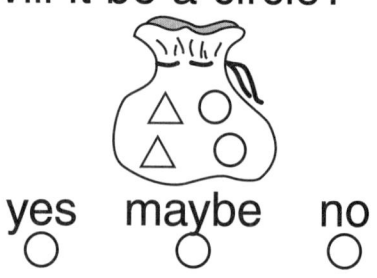

yes   maybe   no

13. Is the spinner fair?

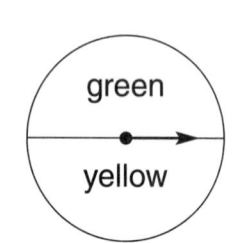

○ yes
○ no

14. Is the spinner fair?

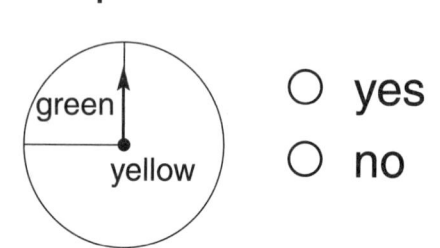

○ yes
○ no

Share the cookies. Draw a picture.

			leftovers
15.	6		
16.	7		

10
Sums and Differences to 18

+	0	1	2	3	4	5	6	7	8	9
0										
1										
2										
3										
4											
5										
6										
7										
8										
9										

Name _____

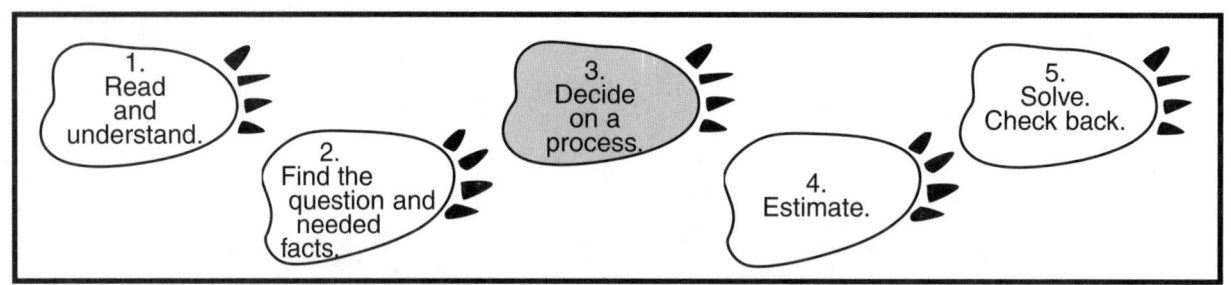

1. Read and understand.
2. Find the question and needed facts.
3. Decide on a process.
4. Estimate.
5. Solve. Check back.

Clue words may help you decide what to do. Ring + or –.

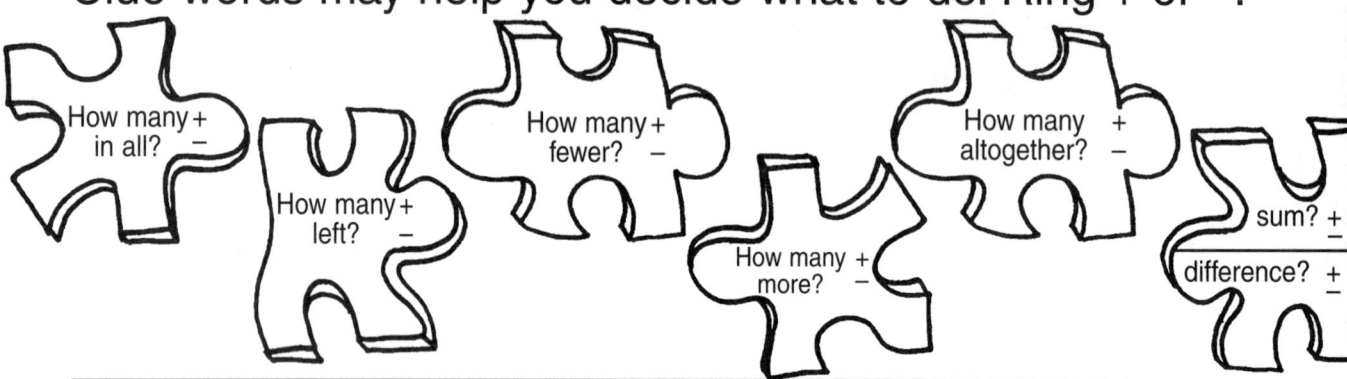

How many in all? + –
How many left? + –
How many fewer? + –
How many more? + –
How many altogether? + –
sum? + –
difference? + –

Underline the clue words in each problem. Ring how to solve.

1. Don saw 6.
He saw 5 more.
How many did he see altogether?

   add   subtract

2. Kim had 12.
She gave 4 away.
How many did she have left?

   add   subtract

3. Pedro bought 8.
Don bought 3.
Kim bought 5.
How many fewer does Don have than Pedro?

   add   subtract

4. Meg got 4.
Pat got 8.
How many did they get in all?

   add   subtract

5. Kim saw 11.
Pedro saw 8.
How many more did Kim see than Pedro?

   add   subtract

6. Don bought 11.
He lost 5.
How many are left?

   add   subtract

Name _____

Reasonable Answers

1. Read and understand.
2. Find the question and needed facts.
3. Decide on a process.
4. Estimate.
5. Solve. Check back.

Don saw 7.
He saw 5 more.
How many did he see?

2   12

One of the answers is not reasonable. Why?

*The answer is more than 7. 2 is **not** reasonable.*

2   (12)

Guess. Ring the correct answer.

1. Kim bought 8.
   3 broke.
   How many does Kim have?

   5   11

2. Pedro had 9.
   He bought 2 more.
   How many does he have?

   7   11

3. Pat had 7.
   Kim had 5.
   Pat ate 2 of his.
   How many does Pat have?

   9   5

4. Pedro had 10.
   He gave away 2.
   How many does he have?

   12   8

5. Meg bought 9.
   Don bought 3.
   How many more did Meg buy?

   6   12

6. Pat had 8.
   Kim had 4.
   How many did Pat and Kim have?

   12   4

Name _____

Many things come in twos.

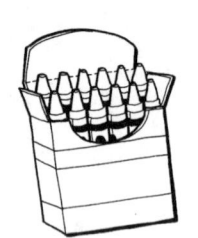

Doubles	Doubles plus 1.
8 + 8 = _____	6 + 7 = _____

1.   8   4   7   8   4   5
      +8   +3   +7   +7   +5   +7

2.   6   8   4   6   7   5
      +6   +9   +5   +8   +6   +5

3.   3   9   6   5   8   3
      +3   +7   +7   +4   +8   +4

4.   9   6   9   7   8   7
      +8   +5   +9   +9   +6   +8

5.   7   4   7   7   8   5
      +8   +4   +5   +7   +7   +6

© Math Teachers Press, Inc. Reproduction by any means is strictly prohibited.

Mental Math: Hungry Bug 9

Meet the Hungry Bug. It always needs 10 cubes to be full.

How many of these 3 cubes must the Hungry Bug 9 eat to be full?

The Hungry Bug 9 eats 1 of the 3 cubes to be a full 10.

9 + 3 is the same as 10 + 2 or 12

1. 9   3   9   2   9   1
   +3   +9   +1   +9   +2   +9

2. 4   6   9   2   9   8
   +9   +9   +3   +9   +4   +9

3. 9   9   9   7   7   5
   +6   +5   +7   +8   +9   +9

4. 6   9   9   8   9   6
   +9   +9   +8   +5   +9   +8

5. 9 + 7 = _____   5 + 9 = _____   8 + 9 = _____

6. 9 + 9 = _____   9 + 6 = _____   4 + 9 = _____

Name _____

Addition Table

Write the sums.

+	0	1	2	3	4	5	6	7	8	9
0										
1										
2										
3										
4										
5										
6										
7										
8										
9										

Name _____

13 bears are on the plane.
4 bears get off.
How many bears are on the plane?

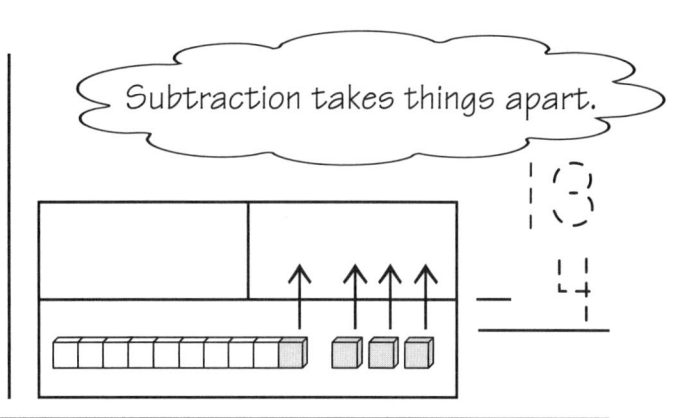

> Subtraction takes things apart.

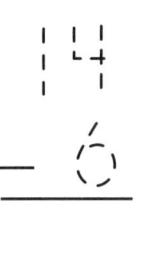

13
− 4
___
 9

Use blocks or counters to subtract.

1. 14 bears are on the plane.
   6 bears get off.
   How many bears are on the plane?

   14
   − 6
   ___

2. 13 bears are on the plane.
   5 bears get off.
   How many bears are on the plane?

Mark Xs on the squares to subtract. Record the answer.

3. 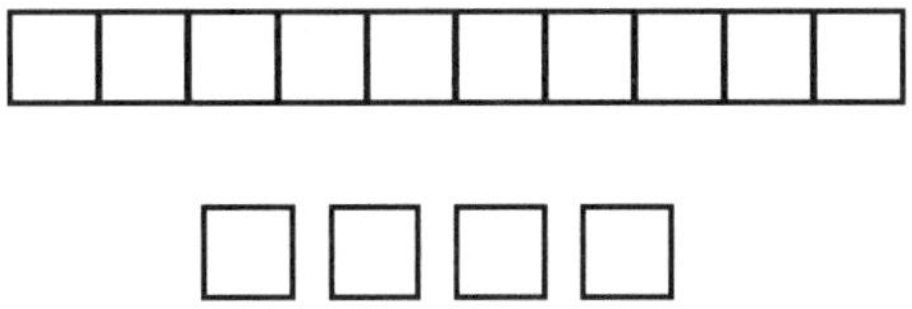 14 − 5 = ____

4. 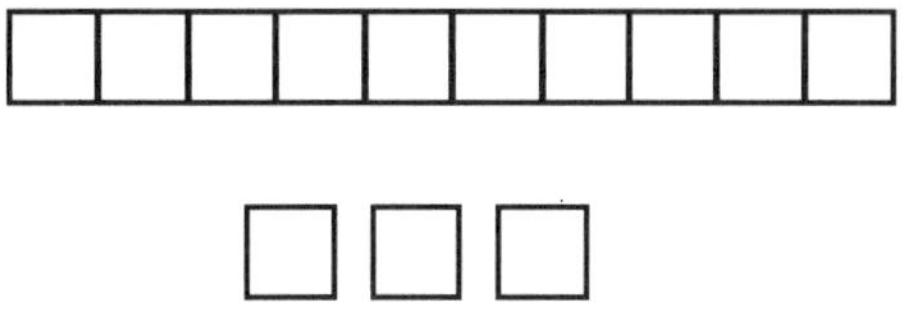 13 − 6 = ____

Use blocks or your mat to subtract.

5. 13   14   13   14   13   14
   − 7   − 7   − 9   − 8   − 8   − 9
   ___   ___   ___   ___   ___   ___

6. 13   14   13   14   13   14
   − 5   − 5   − 6   − 6   − 4   − 7
   ___   ___   ___   ___   ___   ___

Name _____

Kim and Pat are playing a skipping game to practice subtraction. To find 16 − 9, Pat skipped forward 16 times, turned around and skipped back 9 times. Where did he land?

Start at 16. Skip back 9 times.
*15, 14, 13, 12, 11, 10, 9, 8, **7**.*

16
− 9

0 1 2 3 4 5 6 7 8 9 10 11 12 13 14 15 16 17 18

Use your finger or a pencil to skip forward and back on the number line.

1. 15   16   17   18   15   16
   − 6   − 8   − 8   − 9   − 8   − 7

2. 16   15   17   14   13   15
   − 9   − 7   − 9   − 6   − 4   − 9

Use your paper counting strips to subtract.

3. 17   15   16   13   18   15
   − 9   − 6   − 9   − 6   − 9   − 8

4. 16   13   16   15   13   14
   − 7   − 5   − 8   − 7   − 8   − 7

5. 13   14   13   17   14   13
   − 7   − 5   − 9   − 8   − 8   − 4

Name _____

Relating Addition and Subtraction

9 bears are on the plane.
4 more bears get on.
How many bears are
on the plane?

$\begin{array}{r} 9 \\ +4 \\ \hline \end{array}$

13 bears are on the
plane. 4 bears get off.
How many bears are
on the plane?

$\begin{array}{r} 13 \\ -4 \\ \hline \end{array}$

(Draw a picture of 4 more bears)

(Mark an X on 4 bears)

Addition puts things together. Subtraction takes things apart.
Each addition fact has 2 related subtraction facts.

If 9 + 4 = 13, then 13 − 4 = ☐ and 13 − 9 = ☐

Use your blocks.
Write 2 subtraction facts for each addition fact.

1. 6 + 7 = 13   13   13
    −___ −___

2. 7 + 8 = 15   15   15
    −___ −___

3. 9   15   15
   + 6   −___   −___

4. 6
   + 8   −___   −___

5. 7
   + 9   −___   −___

6. 5
   + 9   −___   −___

Missing Numbers ?

How many are hiding?

There are 7 cubes. ▫▫▫▫▫▫▫	You hide some cubes under a paper. How many are hiding? ▫▫▫▫ ?	$4 + \boxed{?} = 7$ *4 + 3 more cubes is 7.* $\boxed{?} = 3$
1. You start with 8 cubes. ▫▫▫▫▫▫▫▫	You hide some cubes. How many are hiding? ▫▫▫▫▫▫ ?	$6 + \boxed{?} = 8$ $\boxed{?} =$
2. You start with 9 cubes. ▫▫▫▫▫▫▫▫▫	You hide some cubes. How many are hiding? ? ▫▫▫▫	$\boxed{?} + 5 = 9$ $\boxed{?} =$

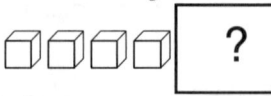

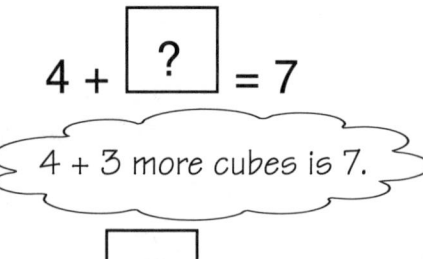

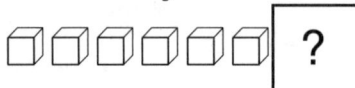

3. There are 8 🐰.
   You see 3 🐰.
   How many are hiding?

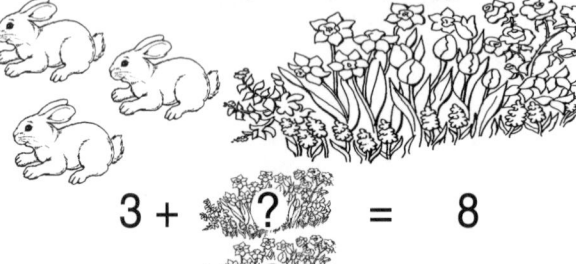

   $3 + \boxed{?} = 8$
   $\boxed{?} = \underline{}$

4. There are 10 🐰.
   You see 2 🐰.
   How many are hiding?

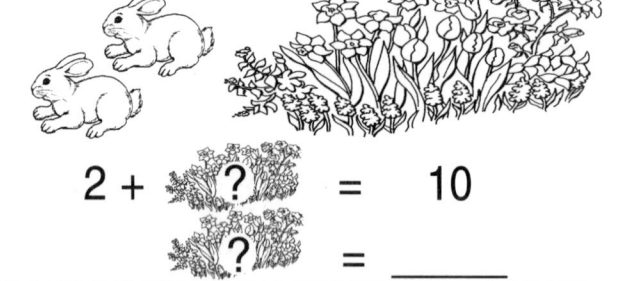

   $2 + \boxed{?} = 10$
   $\boxed{?} = \underline{}$

5. $4 + \boxed{?} = 10$ $4 + \boxed{?} = 7$ $9 + \boxed{?} = 11$
        $\boxed{?} = \underline{}$ $\boxed{?} = \underline{}$ $\boxed{?} = \underline{}$

6. $7 + \boxed{?} = 9$ $\boxed{?} + 8 = 11$ $6 + \boxed{?} = 11$
        $\boxed{?} = \underline{}$ $\boxed{?} = \underline{}$ $\boxed{?} = \underline{}$

© Math Teachers Press, Inc. Reproduction by any means is strictly prohibited.

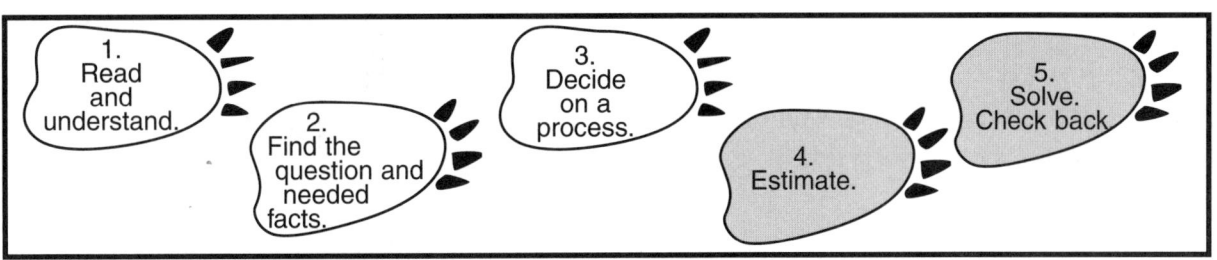

Name _____

1. Read and understand.
2. Find the question and needed facts.
3. Decide on a process.
4. Estimate.
5. Solve. Check back.

1. Meg bought 8 🎟.
   Don bought 3 🎟.
   Each ticket cost 10¢.
   How many 🎟 did they buy?

   Estimate 10   11

   $\begin{array}{r} 8 \\ + 3 \\ \hline 11 \end{array}$

2. Pedro won 15 prizes.
   Kim won 9 prizes.
   How many more prizes did Pedro win?

   Estimate _____   _____

3. Pat saw 7 🤡.
   Meg saw 9 🤡.
   How many 🤡 did they see in all? _____

   Estimate _____   _____

4. Don had 17 rides.
   Pedro had 8 rides.
   How many more rides did Don have? _____

   Estimate _____   _____

5. Kim won 3 🧸.
   Meg won 8 🧸.
   How many 🧸 did they win in all? _____

   Estimate _____   _____

6. Pat and Don saw 5 horses and 9 goats at the Children's Farm. How many animals did they see? _____

   Estimate _____   _____

7. Don bought 15 🎟 and gave away 9. How many 🎟 did he have left? _____

   Estimate _____   _____

8. There are 4 red seats, 7 green seats and 5 yellow seats on the Ferris Wheel. How many seats are on the Ferris Wheel? _____

   Estimate _____   _____

249

Name _____

Fact Families

Use the number given.
Complete the number sentences.

1. **14** / 9 5

   _9_ + _5_ = 14   _14_ − _9_ = 5
   _5_ + _9_ = 14   _14_ − _5_ = 9

2. **15** / 8 7

   ___ + ___ = 15   ___ − ___ = 8
   ___ + ___ = 15   ___ − ___ = 7

3. **16** / 9 7

   ___ + ___ = 16   ___ − ___ = 9
   ___ + ___ = 16   ___ − ___ = 7

4. **17** / 9 8

   ___ + ___ = 17   ___ − ___ = 8
   ___ + ___ = 17   ___ − ___ = 9

Name _____

Problem Solving: Guess and Check

balloon 7¢   ticket 5¢   juice 6¢   popcorn 8¢   bear 9¢

Look at the pictures. Write a number sentence to answer each question.

1. Meg bought a 🍿 and a 🥤.
   How much did she spend?

   8
   + 6
   ___
   ___ ¢

2. Don had 15¢.
   He bought a 🥤.
   How much did he have left?

   ___ ¢

3. Pedro spent 15¢.
   He bought 🍿 for 8¢.
   What else did he buy?

   ___ ¢
   _____

4. Pat has 15¢.
   She wants to buy 2 tickets and a juice.
   Does she have enough money?   ___ ¢

   Yes   No

5. Meg bought a 🐻.
   Jane bought a 🎈.
   How much more did Meg spend?

   ___ ¢

6. Kim spent 16¢.
   She bought a balloon for 7¢. What else did she buy?

   ___ ¢
   _____

7. Dick bought a 🎟, a 🥤 and a 🎈.
   How much did he spend?

   ___ ¢

8. Jack bought 2 different items. He spent 12¢.
   What did Jack buy?

   _____ and ___ ¢
   _____

Names for a Number

Ring addition names for each number.

1.	8	(4 + 4)	6 + 3	(5 + 3)	2 + 5	(2 + 6)
2.	5	3 + 3	2 + 3	0 + 5	5 + 1	4 + 1
3.	11	4 + 7	9 + 2	7 + 3	6 + 6	8 + 3
4.	10	2 + 9	9 + 1	3 + 7	6 + 4	4 + 5
5.	6	4 + 2	1 + 4	4 + 3	3 + 3	1 + 5
6.	9	6 + 4	3 + 5	7 + 2	4 + 5	8 + 1

Ring subtraction names for each number.

7.	5	(12 − 7)	15 − 8	12 − 6	(11 − 6)	(14 − 9)
8.	8	15 − 6	17 − 9	14 − 7	16 − 8	13 − 5
9.	4	10 − 6	13 − 9	9 − 4	12 − 5	11 − 7
10.	6	12 − 6	18 − 9	15 − 7	14 − 8	13 − 7
11.	9	17 − 8	15 − 7	14 − 5	18 − 9	16 − 8
12.	3	10 − 7	11 − 8	9 − 5	12 − 9	8 − 6

Adding 3 Numbers

Kim bought 5 tickets.
Pedro bought 4 tickets.
Don bought 3 tickets.
How many tickets did they buy in all?

$$\begin{array}{r} 5 \\ 4 \\ +3 \\ \hline \end{array}$$

1.
```
   3        0        8        4        2        3
   5        7        8        3        5        1
  +4       +2       +2       +7       +8       +7
```

2.
```
   4        8        5        6        2        3
   2        1        4        3        2        1
  +6       +5       +4       +2       +4       +7
```

Add. Look for combinations of 10.

3.
```
   2        3        1        3        7        3
   5        1        7        3        5        2
  +8       +9       +0       +7       +3       +7
```

4.
```
   9        8        5        2        2        6
   0        4        1        1        5        5
  +6       +2       +5       +7       +6       +4
```

253

Name _____

Add or subtract.

Watch the signs.

1. 6   8   9   7   8   5
   +7   +3   +7   +7   +6   +9

2. 13   16   12   11   14   17
   − 5   − 8   − 9   − 7   − 6   − 8

Watch out!

3. 5   13   18   5   9   12
   +8   − 4   − 9   +6   +4   − 5

Write + or − in each box.

4. 3 ☐ 4 = 7   6 ☐ 7 = 13

5. 13 ☐ 6 = 7   12 ☐ 4 = 8

6. 16 ☐ 7 = 9   14 ☐ 9 = 5

7. 7 ☐ 2 = 9   8 ☐ 9 = 17

8. 11 ☐ 7 = 4   8 ☐ 8 = 16

254
© Math Teachers Press, Inc. Reproduction by any means is strictly prohibited.

Name _____

Chapter 10 Review

1. 8 bears are on the plane. 4 more bears get on. How many bears are on the plane?
(Draw a picture.)

   8
 + 4

2. 12 bears are on the plane. 4 bears get off. How many bears are on the plane?
(Cross out the bears getting off the plane.)

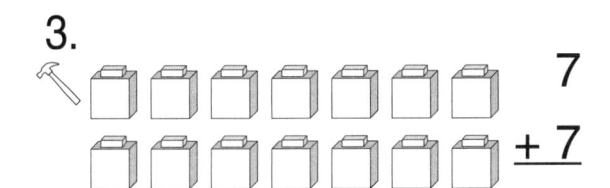

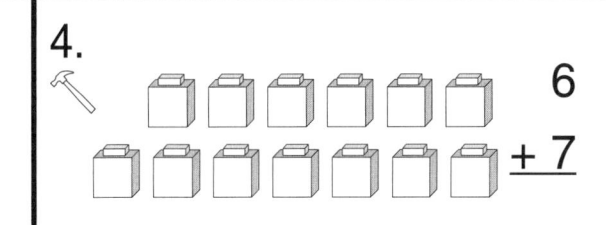

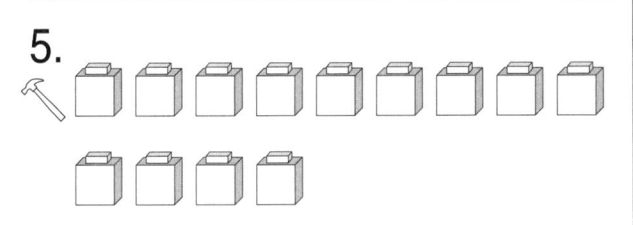

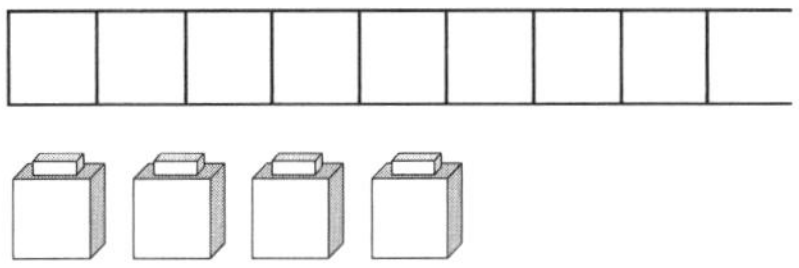

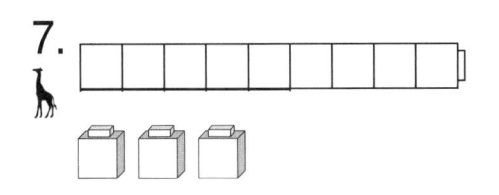

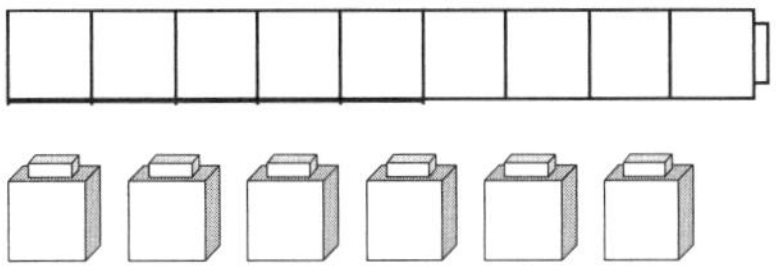

   12
 − 4

3.   7
   + 7

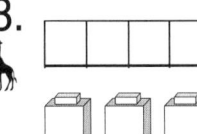

4.   6
   + 7

5. 9 + 4 = _____

6.   14
   − 5

7.   13
   − 6

8.   16
   − 8

Name _____

9. Ring how to solve.
Don saw 7 🐑.
He saw 4 more 🐑.
How many did he see altogether?

   add   subtract

10. Ring how to solve.
Kim had 11 🎈.
She gave away 3.
How many did she have left?

   add   subtract

11. Guess. Ring the best estimate.
Juan bought 8 🎈.
5 🎈 broke.
How many does Juan have?

   5   15

12. Guess. Ring the best estimate.
Meg had 8 🎟.
Meg bought 4 more 🎟.
How many does she have?

   5   15

13. Ring another name for the number in the box.

   ⬜ 9

   6+4   3+5   7+2   8+0

14. Ring another name for the number in the box.

   ⬜ 4

   10-6   13-8   9–4   9–6

15. Add.

       2
       5
   + 8
   ———

16. Write + or – in each box.

   9 ☐ 2 = 7

   6 ☐ 6 = 12